Out of the Lab and on the Market

How Sony Computer Science Labs (Sony CSL) Turn Research into Profits

Out of the Lab and on the Market

How Sony Computer Science Labs (Sony CSL) Turn Research into Profits

Tetsu Natsume
Mario Tokoro

CRC Press
Taylor & Francis Group
Boca Raton London New York

CRC Press is an imprint of the
Taylor & Francis Group, an **informa** business

A PRODUCTIVITY PRESS BOOK

CRC Press
Taylor & Francis Group
6000 Broken Sound Parkway NW, Suite 300
Boca Raton, FL 33487-2742

© 2018 by Tetsu Natsume and Mario Tokoro
CRC Press is an imprint of Taylor & Francis Group, an Informa business

No claim to original U.S. Government works

Printed on acid-free paper

International Standard Book Number-13: 978-1-138-73590-3 (Hardback)

This book contains information obtained from authentic and highly regarded sources. Reasonable efforts have been made to publish reliable data and information, but the author and publisher cannot assume responsibility for the validity of all materials or the consequences of their use. The authors and publishers have attempted to trace the copyright holders of all material reproduced in this publication and apologize to copyright holders if permission to publish in this form has not been obtained. If any copyright material has not been acknowledged, please write and let us know so we may rectify in any future reprint.

Except as permitted under U.S. Copyright Law, no part of this book may be reprinted, reproduced, transmitted, or utilized in any form by any electronic, mechanical, or other means, now known or hereafter invented, including photocopying, microfilming, and recording, or in any information storage or retrieval system, without written permission from the publishers.

For permission to photocopy or use material electronically from this work, please access www.copyright.com (http://www.copyright.com/) or contact the Copyright Clearance Center, Inc. (CCC), 222 Rosewood Drive, Danvers, MA 01923, 978-750-8400. CCC is a not-for-profit organization that provides licenses and registration for a variety of users. For organizations that have been granted a photocopy license by the CCC, a separate system of payment has been arranged.

Trademark Notice: Product or corporate names may be trademarks or registered trademarks and are used only for identification and explanation without intent to infringe.

Library of Congress Cataloging-in-Publication Data

Names: Natsume, Tetsu, author. | Tokoro, Mario, 1947- author.
Title: Out of the lab and on the market : how Sony Computer Science Labs (Sony CSL) turn research into profits / Tetsu Natsume and Mario Tokoro.
Description: Boca Raton, FL : CRC Press, 2018.
Identifiers: LCCN 2017006695 | ISBN 9781138735903 (hardback : alk. paper)
Subjects: LCSH: Sony Computer Science Laboratories. | Research, Industrial. | Technological innovations. | New products.
Classification: LCC T178.S66 N37 2018 | DDC 338.7/621390952--dc23
LC record available at https://lccn.loc.gov/2017006695

Visit the Taylor & Francis Web site at
http://www.taylorandfrancis.com

and the CRC Press Web site at
http://www.crcpress.com

Printed and bound in Great Britain by
TJ International Ltd, Padstow, Cornwall

Contents

Preface xi
Authors xiii

SECTION I How Sony CSL Turns Research into Profits

Chapter 1 Birth of the Technology Promotion Office 3

    1.1   The Daily Grind of a Technology Promoter 3
    1.2   Technology Promotion: Navigating the Gaps of Timing and Setting between Business and Research 7
        1.2.1   The Invention of Augmented Reality in the Late 1990s 7
        1.2.2   Thirteen Years of AR Development Comes to Fruition in a Video Game 9
        1.2.3   Navigating Gaps in Timing and Setting to Bring Research to Market 11
    1.3   What Is the TPO? 12
        1.3.1   Meeting Mario 13
    1.4   Assembling a Corps of Researchers with Ideas Just Crazy Enough to Change the World 14
        1.4.1   Vision of the Ideal Research Laboratory: Contributing to Humanity and Society as Well as SONY 14
        1.4.2   Senior Researchers Carving Out New Fields of Study 15
        1.4.3   Blossoming of Youthful Talent 17
    1.5   Sony CSL-Derived Technology Powering Xperia™ Smartphones 18
        1.5.1   POBox Predictive Text Input, and One-Touch Devices Based on FEEL 18
        1.5.2   Making Devices More Intuitive: Lifelog and Smart Operation Gesture Recognition 20
        1.5.3   Sony CSL Is an Outlier among SONY's Many R&D Organizations 21

- 1.6 Mired in Obscurity 23
  - 1.6.1 Products Carrying CSL's Technology DNA, but Not CSL's Name 23
  - 1.6.2 Changing Sony CSL's Research Style 24
- 1.7 TPO Takes Flight 25
  - 1.7.1 Mission: Maximize the Fruits of CSL Research 25
  - 1.7.2 Three-Point Action Plan 26
    - 1.7.2.1 Point 1: Take Inventory of Sony CSL Research. What Did That Entail? 27
    - 1.7.2.2 What about Point 2: Lock Our Sights on the Most Promising Targets? 27
    - 1.7.2.3 And Finally Point 3: Maximize Impact 27

Chapter 2 Case Studies in Technology Transfer 29
- 2.1 VAIO Pocket: A Painful Learning Experience 29
  - 2.1.1 Most Technology Transfers Go Nowhere 29
  - 2.1.2 Presense Technology and the VAIO Pocket 30
- 2.2 The Difficulty of Taking Technical Breakthroughs to Market 31
  - 2.2.1 FEEL: A Landmark Idea 31
  - 2.2.2 FEEL's Lifeline: A Videoconferencing System 33
  - 2.2.3 Adoption into the NFC Standard 34
  - 2.2.4 FEEL-Enabled Phones Hit the Market and the "MoTR" Saga 35
  - 2.2.5 The Development of One-Touch 36
- 2.3 A CSL Paris Technology's Unexpected Route to Success 38
  - 2.3.1 EDS: Music Categorization Technology from Paris 38
  - 2.3.2 Building a Win–Win Relationship with SONY R&D 39

	2.4	Interindustry Collaborations 40
		2.4.1 Moe-Kaden: Giving Digital Appliances a Human Face 40
		2.4.2 Barnstorming Negotiations with Daiwa House and within SONY 42
	2.5	Going to Market with a Product Targeting Teenage Girls 43
		2.5.1 An App That Broke the Galapagos Barrier ... 43
		2.5.2 Proving the Power of a Public Beta 45

Chapter 3 Next-Level Challenges for the Technology Promoter 49

	3.1	Econophysics Optimizes Semiconductor Production 49
		3.1.1 It's Called Econophysics: Now Where Can We Use It? 49
		3.1.2 Forging a Unique Partnership with the Semiconductor Business Group 51
		3.1.3 A Top-Shelf Example of Bottom-Line Gains 52
	3.2	A New Outlet for Research Discoveries: Science Content for Entertainment Media 53
		3.2.1 Ken Mogi's "Aha! Experience" Draws Attention and Sega Wants In 53
		3.2.2 Tapping Sony Music Artists to Do a Deal with Sega 54
		3.2.3 An Aha! Experience Ecosystem with an Eight-Digit Dollar Value 56
	3.3	Sony CSL's First Spin-Off 58
		3.3.1 The Place Project: A New Kind of Location-Sensing Service 58
		3.3.2 Creating a Buyer 60
		3.3.3 Forging a Flagship Product for the Tokyo National Museum 61
	3.4	Toward a New Electric Power Industry 63
		3.4.1 Creating a Next-Generation Electrical Infrastructure 63
		3.4.2 A Large-Scale Project Involving Academic and Corporate Partners 64

		3.4.3	"Packetized Electric Power": A Concept Borrowed from the Internet 65
		3.4.4	Making the Connection with SONY's Battery Biz 65
		3.4.5	Putting on Public Viewings of the World Cup in Ghana 66
		3.4.6	Launching a Mobile Phone Charging Service for Unelectrified Areas 69
		3.4.7	Forming a Consortium for the "Internet of Electricity" 71

Chapter 4 Techniques for Technology Promotion 75
- 4.1 Cataloging of Research Discoveries 75
  - 4.1.1 The *Review Talk* 75
  - 4.1.2 The TPO Interview 76
  - 4.1.3 Multiple Information-Gathering Channels 78
- 4.2 Developing Sales Collateral 80
  - 4.2.1 A Sales Sheet That Zeroes in on "What Does It Do?" 80
  - 4.2.2 Technology Tag = Schematized FAQ 81
- 4.3 Selling 82
  - 4.3.1 Demo Road Show: Empowering TPO to Make Initial Pitches Independently 83
  - 4.3.2 *T-pop News*: An E-Mail Newsletter for People Who Have Attended Our Demos 85
- 4.4 The Importance of the MoTR 88
- 4.5 Philosophy on License Fees for Technology 91
- 4.6 The 10 Core Principles of Technology Promotion 92
  - 4.6.1 Principle 1: There's a Right Time to Bring Every Discovery out of the Lab 94
  - 4.6.2 Principle 2: Use Every Possible Connection 94
  - 4.6.3 Principle 3: Don't Hold Preconceptions About Other Organizations 94
  - 4.6.4 Principle 4: Assume That Customers Need a Simple Message Drilled in Hard 95
  - 4.6.5 Principle 5: Mental Fortitude 95
  - 4.6.6 Principle 6: Try Things That Haven't Been Tried Before 95

	4.6.7	Principle 7: Move Fast! Every Second Counts 95
	4.6.8	Principle 8: Follow-Up After Technology Transfer Is a Must 96
	4.6.9	Principle 9: Don't Just Be a Technology Promoter, Be a Research Producer 96
	4.6.10	Principle 10: Never Forget That You're Doing It for the Lab 96
4.7	Why Technology Promotion Is Necessary 96
	4.7.1	Overcoming Divergences of Timing and Setting 97
	4.7.2	Role Division between Researchers and Technology Promoters 98
	4.7.3	Technology Promotion Must Be Part of the Lab 98
	4.7.4	Beyond Technology Promotion 98
	4.7.5	How Selling Research Changes Research 99

SECTION II Researchers on Technology Promotion

II.1 Implementing Outrageous Ideas 101
Alexis Andre, CSL Researcher

II.2 Implementation of Academic versus Corporate Research 103
Jun Rekimoto, CSL Deputy Director

II.3 Papers Are Fine, But Nothing Beats the Joy of Research That Becomes Products That Change the World! 106
Takashi Isozaki, CSL Researcher

SECTION III The History of Sony's Technology Promotion Office (Mario Tokoro)

Chapter 5 Before TPO 111

5.1 Lab-Driven Product Development: The Precursor to TPO 111

5.2 Jigsaw Puzzles with Pieces Missing and Assets Left to Rot 113

Chapter 6 From the Perspective of Technology Management...... 115

      6.1 Business Management and Innovation 115

      6.2 Technology: Development Process, Time, and Cost.. 116

      6.3 Horizontal Business Models and Open Innovation.. 118

      6.4 Management of Open Innovation............................ 120

Chapter 7 What TPO Represents... 123

      7.1 The Implementation of TPO 123

      7.2 What TPO Represents.. 124

In Closing: Borderless Technology Promotion................................. 127

Afterword 131

Index 133

Preface

"Our research produced great results, but the business division didn't pick up on it, so it had no chance of becoming a product."

"We came up with a new technology, but we don't know who to pitch it to."

"That hot new technology is exactly like something we were working on seven years back. That hurts! If we'd managed to commercialize it at the time, it would have been a global first."

Alongside the effort that goes into the study itself, I am sure that many researchers also encounter frustrations while trying to shepherd their work through to implementation.

And I don't doubt that laboratory managers also spend many a sleepless night fretting over how best to improve the performance of their lab, how best to generate fresh research successes, and how to commercialize that research.

"I don't understand why the business division passed on the chance to implement our lab's research."

"Improving on existing technologies is all well and good, but the Holy Grail for laboratories is something entirely new. But even if we do come up with something like that, there's nowhere for us to take it!"

I feel that one possible solution to such concerns is the introduction to laboratories of a technology promotion team.

My name is Tetsu Natsume. At Sony Computer Science Laboratories (Sony CSL), I am responsible for running a research marketing body named the Technology Promotion Office (TPO). It is now more than a decade since this office began its activities in August 2004.

Just as writers have editors, film directors have producers, and athletes have coaches, in order for talent to bear fruit, creative individuals need a supportive presence close at hand. In short, researchers have a need for technology promotion.

But just as the job of an editor differs from that of a coach, technology promotion also carries its own peculiar responsibilities. Although at the outset, my own vision for TPO was somewhat hazy, after 10 years of hitting countless walls, and much trial and error, I have managed to work out the right approach. And looking back on the last decade, I have come to see that all our activities to date represent key aspects of technology promotion.

Through this book, I hope readers will get a sense of the activities and techniques associated with technology promotion, which I see as indispensable for the laboratories of the future.

Sections I and II of this volume were written by me. Section III, was penned by Mario Tokoro, founder of Sony CSL. He explains the meaning of technology promotion from the perspective of technology management, and I feel that the numerous examples of research marketing from our own laboratory also provide a model for the future running of laboratories involved in fundamental research.

It is my hope that this book can promote the implementation of such activities around the world. And I also hope that by fostering closer connections between research and business, we can contribute at least a little to an increase in the successful utilization of research results for the benefit of society and of humanity.

Representing the authors

Tetsu Natsume
Head of Technology Promotion Office
Senior General Manager
Sony Computer Science Laboratories

Authors

Tetsu Natsume, B.Sc., After graduating from the Department of Geology (major in Geography) in the Faculty of Science at the University of Tokyo, Tetsu Natsume joined SONY Corporation in 1988. Engaging in various activities relating to electronic devices, including production management, business management, and legal administration of exports and imports, he spent six years in Singapore on two assignments—at a local factory and the regional headquarters. He then spent some time with the Business Strategy Division at SONY Headquarters before engaging in the business development of a 360-degree, unrestricted viewpoint video system. This experience drove home to him the necessity of a technology promotion body that could handle the practical implementation of multiple research topics. In 2004, he was transferred to Sony Computer Science Laboratories, Inc., where he launched and continues to run the Technology Promotion Office (TPO). In the intervening years, he has been involved in numerous forms of research commercialization and the establishment of spin-off ventures.

Dr. Mario Tokoro, Ph.D., is Founder and Executive Advisor, Sony Computer Science Laboratories, Inc. (http://www.sonycsl.co.jp/en). Dr. Tokoro was Professor of Computer Science at Keio University when he established Sony Computer Science Laboratories, Inc., in 1988 and went on to guide its development into a world-renowned fundamental research laboratory. In 1997, he moved to SONY Corporation to become Corporate Senior Vice President, and then in 2000, Chief Technology Officer.

Dr. Tokoro proposed and has been promoting a new scientific methodology called Open Systems Science to address challenges arising from complex, ever-changing systems such as earth sustainability, life and health, and huge man-made information infrastructures (*Open Systems Science: From Understanding Principles to Solving Problems*, IOS Press, 2010). Applying this methodology, he developed a new software process called DEOS (*Open Systems Dependability: Dependability Engineering for Ever Changing Systems*, 2nd Edition, CRC Press, 2015). He is currently leading the Open Energy Project at Sony CSL.

Dr. Tokoro received *Officier de l'Ordre National du Merit* from the Republic of France in 2005 and *Docteur Honoris Causa* from the University of Paris (UPMC) in 2010.

Section I

How Sony CSL Turns Research into Profits

1

Birth of the Technology Promotion Office

1.1 THE DAILY GRIND OF A TECHNOLOGY PROMOTER

Let me give you a picture of what it looks like to be on the sales force of a research organization by walking you through some typical workday scenarios for a staff member of Sony Computer Science Lab's (CSL) Technology Promotion Office (TPO).

> **A RESEARCH LABORATORY AT SONY CSL—DAYTIME**
>
> TETSU NATSUME and RESEARCHER "A" are in animated conversation.
>
> **TETSU**
> *(rubbing his temples)*
> I wasn't able to follow your recent
> briefing to HIROAKI about
> your research progress.
> Could you go over it again and put
> it in layman's terms for me?
>
> **RESEARCHER "A"**
> *(sighing)*
> You're kidding me...
> Fine, fine, I'll just have to go
> over it with you again.

Researcher "A" goes through a PowerPoint deck of fiendishly technical slides describing his research.

TETSU

(*with resolve*)
It's a tricky technology to explain,
isn't it? I need to put together some
materials that will enable
non-specialists to get the gist of it.
You'll check my draft, right?
And can I use the graphics from
your slide deck?

(*musing*)
And we need to give your technology
a catchy name that conveys what's
unique about it.
I'm not going to be able to sleep at
night until we hit on the right name.
So help me brainstorm!

[SOME TIME LATER]

A CONFERENCE ROOM AT SONY HQ

TETSU is trying to pitch the aforementioned technology to a bunch of product engineers.

ENGINEER NO. 1

(*intrigued*)
I had no idea anybody had come up
with a technology that can do this!
What's the latency?

TETSU

Very low latency, from what I
understand. I can get exact figures
from the researcher...

ENGINEER NO. 2

How many lines of code is it?

ENGINEER NO. 3

What makes this any different from the PRODUCT Z technology that COMPANY Z has out there...?

ENGINEER NO. 2

Could it handle a dataset like...
(*starts to spew
impenetrable jargon*)

TETSU

(*beads of sweat
break out on forehead*)
Uhh...

ENGINEER NO. 1

(*in exasperation*)
If you can't answer any of the technical questions we need answered in order to evaluate this technology, what are you doing here? Get back to us when you understand what you're pitching!

SONY CSL RESEARCH LABORATORY

TETSU and RESEARCHER "A" debrief.

TETSU

(*with chagrin*)
I was over at BBB Division presenting your technology and they bombarded me with so

many questions, I had to beat
a hasty retreat.
How am I supposed to deal with
an avalanche of technical
questions next time I go?
You have to prep me!

RESEARCHER "A"

(*sighing*)
No problem. I'll go over the
answers to those questions
with you for next time.
Are they going to even give
you a next time?

Researcher "A" goes through a PowerPoint deck of fiendishly technical slides describing his research.

TETSU

(*clenches fist*)
I'm not giving up! I'm going
back in there to pitch again!
And if I don't make any headway,
I'm going to bypass MANAGER "X".
He's a tough nut to crack.
I bet MANAGER "Y" would be
an easier sell on this.

TETSU'S OFFICE

His mobile phone rings.
Glancing down at the screen to see who it is, he quickly takes the
call and exchanges greetings.

MANAGER "Q"

We're plotting out our product
lineup for the year after next

and I'm wondering if you have
any interesting technologies
brewing over at CSL.
Our planning team is meeting.
Got anything to demo for us?

TETSU

I sure do! And I'll come running
to show it off to you at the
drop of a hat!
How about this evening?

Tetsu pumps his fist.

Scenes like this play out every day in the TPO as we pitch the innovations that bubble up from our pool of ingenious researchers to SONY corporate, SONY-affiliated companies, and beyond.

Maybe you're wondering why a research organization needs to have a sales office like ours. Let me walk you through how our organization is set up and what it does.

1.2 TECHNOLOGY PROMOTION: NAVIGATING THE GAPS OF TIMING AND SETTING BETWEEN BUSINESS AND RESEARCH

1.2.1 The Invention of Augmented Reality in the Late 1990s

The standard pattern in the business world is for corporate research units to develop technologies and then transfer them to business units, which turn them into products and launch them into the marketplace. But Sony CSL engages in blue-sky research, aiming to conjure up breakthrough technologies that are 5 or 10 years ahead of their time. This often means the marketplace isn't ready for them when they first surface. In other words, there is a timing gap between the activities of research and business organizations.

Adding to that, one trait of the kind of minds who do blue-sky research is that they are always thinking up new things, things the world has never seen before. Ask one of them what they've been up to recently, and they will want to talk about their very latest findings and achievements, fresh from the lab. They aren't interested in talking about research they worked on a decade ago. But in many cases, what the marketplace is finally ready for *now* is exactly those technologies that they worked on 10 years ago!

That's where technology promotion comes in. As researchers focus on pushing the envelope of cutting-edge discoveries, we serve as a research sales force relentlessly chasing down opportunities to commercialize research findings.

A case study I'd like to share with you concerns augmented reality (AR) technology. AR refers to the display of virtual objects or information generated by computer graphics superimposed on the display of real-world environments, allowing people to more intuitively access information and to interact with virtual objects. This field was pioneered in the early 1990s by Jun Rekimoto, who currently serves as Sony CSL's deputy director (Figure 1.1).

Cast your mind back to the early 1990s and recall that the Internet had not yet become the foundation of all business and social life; merely hooking up a camera to a computer remained a tricky proposition, remember? The state of computer graphics back then . . . well, the original PlayStation™ (PS) had not even been released. Only pricey workstations had the horsepower to display even primitive graphics, which makes it really quite astonishing that anyone was undertaking the research that Jun was then. And as for commercializing AR? Out of the question given the hardware on the market at the time.

But by the turn of the millennium, digital cameras had become small and cheap. By then, not only personal computers (PCs) but even mobile

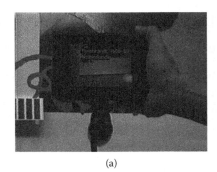
(a)

(b)

FIGURE 1.1

(a, b) NaviCam, an early AR research project by Jun Rekimoto. (Courtesy of Sony Computer Science Laboratories, Inc., Tokyo, Japan.)

phones routinely came with cameras built in. And on the computer graphics front, devices like the PS and smartphones had already become capable of deftly generating compelling images. In other words, the time had become ripe for a technology like AR.

1.2.2 Thirteen Years of AR Development Comes to Fruition in a Video Game

One of the AR technologies Jun developed is called CyberCode. CyberCode was a groundbreaking, next-generation barcode technology implementing AR for visual information tagging. It broke open a huge new market space, into which other similar software applications flooded; but after being built into one model of Sony PC, the VAIO-C1, in the 1990s, CyberCode itself was not used continuously. The VAIO with CyberCode generated a huge amount of buzz, being hailed as a groundbreaking technology, but marketplace conditions weren't favorable for it to catch on.

Tsukasa Yoshimura, a veteran of research sales who currently serves as an advisor to the TPO, doggedly pursued commercialization opportunities for CyberCode.

He had been previously involved in a virtual reality (VR) project, which is how he encountered CyberCode. Even in 2003, when his work on the project lay more than 5 years in the past, he continued to toy with its commercial possibilities in the back of his mind. His work in 2003 brought him into frequent contact with video game producers at Sony Computer Entertainment (SCE). One of them was Satoru Miyaki, and together he and Tsukasa hatched a radically inventive concept for a card battle game on the PS2 that would make use of CyberCode.

The PS2 supported an optional camera accessory called the EyeToy, which enjoyed a strong following, particularly in Europe. The EyeToy was used to read the CyberCode off a physical battling card, and then the PS2 superimposed graphics of the monster it represented on the video of the actual card using AR. When I first heard the idea, I was blown away and helped them make a demo video of it.

But that was where things started to get tricky as we embarked on a long saga of trying to move the AR game forward—with Yuji Ayatsuka, one of the original TPO staff, shepherding the technology side of things while I drove the business side. On the SCE side, executive producer Kazuhito Miyaki (no relation to Akira) and producer Yusuke Watanabe were involved. We had who knows how many meetings.

10 • *Out of the Lab and on the Market*

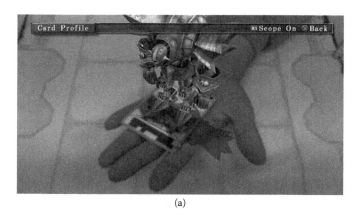

(a)

(b)

FIGURE 1.2
(a, b) PlayStation™ game making use of CyberCode. (Courtesy of THE EYE OF JUDGMENT™ © 2007 Sony Interactive Entertainment, Inc., San Mateo, CA.)

It turned out that video games had a prolonged gestation period, including a number of shifts in plans along the way; therefore, even though the idea germinated in 2003, it was not until a long 4 years later, in 2007, that THE EYE OF JUDGMENT™ was released for the PS3. Tracing back to when Jun originally started his AR research, it was 13 years before it showed up on the market in that game (Figure 1.2).

The game was marketed globally and, as one of the most unique titles of the first wave of PS3 games, sold especially well in Europe. It was honored with the Future Division Award at the 2006 Japan Game Awards (having been announced a year before it hit the shelves) as a product holding great promise for the future. Since that pioneering release, games with AR have continued to appear; so THE EYE OF JUDGMENT can be considered the harbinger of long-cocooned AR technology finally bursting forth to spread its commercial wings.

1.2.3 Navigating Gaps in Timing and Setting to Bring Research to Market

Following THE EYE OF JUDGMENT, AR-enabled games became an established genre on the PS line, best received in Europe, and additional titles continued to hit the market. Meanwhile, another market of AR-enabled advertising was being established. Later, via Koozyt, AR would be adopted in campaigns by pro baseball, energy drinks, and others. With the spread of smartphones, that kind of advertising became a mainstay application (Figure 1.3).

The inventor of this technology, Jun Rekimoto, never imagined when he was doing his initial research that it would end up being used in a card battle game on a game console. Changes in the market landscape and era opened a door into the unanticipated domain of video games enabling the technology to achieve commercialization (Figure 1.4). However, to reach the market via a roundabout path like that requires endurance to stay the course and patience to bide time, never losing sight of the technology, until

FIGURE 1.3
Smartphone commercial making use of CyberCode. GnG (GET and GO) is an AR-marketing tool from Koozyt, Inc., Tokyo, Japan.

12 • *Out of the Lab and on the Market*

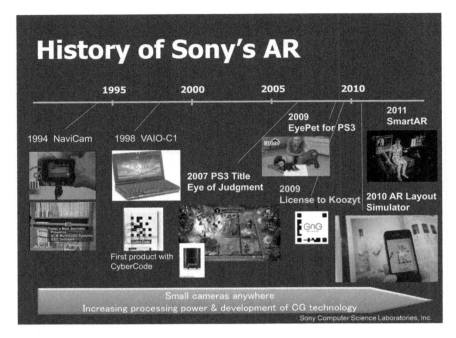

FIGURE 1.4
The progress of AR at Sony. (Courtesy of Sony Computer Science Laboratories, Inc., Tokyo, Japan.)

the right chance comes along—not to mention trying many wrong doors before finding the right one. In other words, research marketing must be advanced steadily through divergences of timing and setting in order to achieve commercialization.

At this point, the timing and setting are right to explain how the TPO is set up organizationally and how it came into existence.

1.3 WHAT IS THE TPO?

The TPO operates out of Sony CSL's Tokyo office. At the time of this writing, it was staffed with three people: myself, Yoko Honjo, and Yoshiichi Tokuda, along with Tsukasa Yoshimura in an advisory role. We are the research sales force for Sony CSL, and our mission is to bring technologies birthed by CSL research to market.

At any given time, the range of research underway is remarkably broad—research that yields novel technologies and discoveries that break new ground in multiple fields. But just because a researcher publishes an academic paper demonstrating new knowledge, that knowledge doesn't automatically lead to a commercial application with real-world value. A process is needed to carry a technology or technique achieved in the lab along the path to being incorporated into an actual product or service in the real world.

Linking the business units that deliver those products and services to the marketplace with the labs that generate the innovations by being a research sales force is the nature of the job of technology promotion. At TPO, we do whatever it takes to usher innovations out of the lab and onto the market.

1.3.1 Meeting Mario

My journey toward eventually taking the reins of Sony CSL's TPO started in 1998, when I had been with SONY for 10 years, during which time I held positions ranging from production management to administration to import/export compliance to government relations. But having majored in the sciences in college, I had always wanted to be involved in launching a business based on new technology. For a long time, though, no opportunities of that nature came my way. In 1998, I happened to be posted to Singapore, where I was involved in government negotiations.

While I was there, it happened that Mario Tokoro, who was a corporate senior vice president at SONY, visited Singapore because he had been named a member of the Scientific Advisory Board of the government's National Research Institute. I was assigned to look after him for 2 days during his stay.

Everything went smoothly over those 2 days; so as I was driving him to the airport for his flight out, I mentioned my hopes of being involved in launching a business connected to VR technology. Mario replied that he had a project going, but that it hadn't reached the commercialization stage yet; he told me to wait 2 years.

Having been posted back to Japan, exactly 2 years later, I went to see Mario at Sony CSL. It was October 11, 2000, the day before my 35th birthday. I had set a goal for myself to be involved in starting up a new business before I hit 35. Mario didn't seem to recall our conversation from 2 years earlier but, true to his word, the project he had mentioned had reached the cusp of commercialization. He set up an interview for me with the project leader, Tsukasa Yoshimura, which took place the next morning—on my 35th birthday.

I gave Tsukasa my best pitch for the business plan that had been simmering in my brain, but when I followed up with him later, he could hardly recall anything about it. What had made an impression on him was me showing up on my 35th birthday determined to make my personal deadline, and he made a snap judgment to put my transfer through. So on January 1, 2001, I joined Tsukasa's project, which was being run out of SONY's corporate research labs under Mario's leadership. It concerned a new 360-degree video technology called FourthVIEW.

The FourthVIEW project was a great learning experience for me, and I would do it all again, but it didn't lead to a successful commercialization. I discovered firsthand how difficult it is to lock in on one particular technology and to try to take that technology, and only that technology, to market. That's when a vague idea began to coalesce in my mind that research divisions needed some kind of supporting function for promoting commercial applications of research, someone who would keep tabs on a raft of technologies and look for opportune moments to promote them. At the time, though, CSL's leaders didn't click with what I was proposing, so after FourthVIEW, I ended up just working in administration there for a while.

When, some time later, Tsukasa came and asked me whether I was interested in doing technology promotion and becoming a salesman for Sony CSL's research, I jumped at it without a second thought.

1.4 ASSEMBLING A CORPS OF RESEARCHERS WITH IDEAS JUST CRAZY ENOUGH TO CHANGE THE WORLD

1.4.1 Vision of the Ideal Research Laboratory: Contributing to Humanity and Society as Well as SONY

Sony CSL is a little bit different from your run-of-the-mill research organization. Corporate research labs tend to be tasked with doing research and development (R&D) on the company's next generation of products. But Sony CSL sees its mission as this: "To contribute extensively to social, industrial, and SONY development through fundamental yet applicable research, especially on and around computer science." SONY is the "also" there—contributing to humanity and society comes first for CSL.

And this is why CSL doesn't confine its research to user interfaces, artificial intelligence (AI), and other fields with obvious connections to SONY businesses but ranges far and wide across biology, neuroscience, economics, medicine, and agricultural science, to name a few. What Sony CSL asks of its researchers above all is not to take the obvious next steps in technological progress but to set their sights on big-time breakthroughs that could change the world. So we bring in the kind of researchers who might just be crazy enough to pull that off.

The basic blueprint came from Sony CSL founder Mario Tokoro, who shaped it to match his 1980s' vision of an ideal research organization. From the outset, he marked Sony CSL as a different breed from typical corporate research labs, one where exceptional researchers would be provided with an environment that empowered them to plunge into research in which they truly believed.

The best way for me to give you a sense of what that looks like is to profile a number of key figures at Sony CSL.

1.4.2 Senior Researchers Carving Out New Fields of Study

Current Sony CSL President and CEO Hiroaki Kitano is a man whose earlier work saw him enjoy great success in the field of AI research. In 1993, his work on AI and speech-to-speech translation won him a Computers and Thought Award, an accolade sometimes referred to as the Nobel Prize of computing. His subsequent involvement with various robot development projects saw him become one of the "parents" of AIBO, SONY's entertainment robot. From the Kitano Symbiotic System Project under the Exploratory Research for Advanced Technology (ERATO) program, many robot ventures—including ZMP and Flower Robotics—were born. He was also one of the founders of the global Robocup robot soccer tournament, and he won renown when he set the target of producing a robot team that could defeat the reigning human soccer World Cup holders by the year 2050. Even while making such strides in the fields of robotics and AI, Kitano has also been active in the biological sciences since the 1990s. His proposal for the new field of systems biology, which incorporates aspects of engineering and information science, has since grown into a significant global research movement.

Deputy Director of Research Jun Rekimoto has also been involved in a wide variety of leading-edge research since the 1990s, becoming one of the key figures in the field of human–computer interaction. One example is AR technology, which involves superimposing computer graphics over the world we see around us and is currently in the midst of a boom on smartphones, tablets, and even glasses and other wearable tech. With a history of AR research going back more than 20 years to the early 1990s, Rekimoto is sometimes referred to as the father of AR. He has also contributed significantly to the latest smartphones through his work in multi-finger, multi-touch operation and other vital features. These days his position at CSL is combined with his role as a professor with the University of Tokyo's Interfaculty Initiatives in Information Studies.

Ken'ichiro Mogi, another CSL researcher, is also a well-known media figure in Japan. With "qualia" (a term coined to express the sensory qualities that accompany conscious perception) as a keyword, Mogi has done extensive research on the relationship between the brain and the mind, while his work on the "aha" experience—along with numerous writings and TV appearances—have made him a household name in Japan. Mogi joined CSL having gained undergraduate degrees from both the Faculty of Science and the Faculty of Law at the University of Tokyo, where he also conducted postgraduate research, as well as spending time at both RIKEN and the University of Cambridge. By combining wide-ranging activities in the fields of neuroscience and social studies, Mogi is credited with raising the profile of neuroscience.

For many years, Kazuhiro Sakurada has been at the forefront of research into the life sciences. Joining Japanese pharma-biotech giant Kyowa Hakko Kogyo straight out of university, he worked on both drug development and research into regenerative medicine. In 2004, he was recruited to head a new research center set up in Kobe by the German pharmaceutical firm Schering. After the company's merger with Bayer to form Bayer Schering Pharma and the combination of the firms' Japanese research divisions, Sakurada's glittering resume came to include such positions as head of the Therapeutic Research Group for Regenerative Medicine, member of the Global Research Management Team, a role on the executive committee of Bayer Yakuhin, and head of the company's Japanese research center. He also gained recognition for his role in Bayer Yakuhin's development of human iPS (induced pluripotent stem) cell technologies, and he is currently focused on research into the fundamentals of human health and longevity.

Hideki Takayasu joined Sony CSL from a professorship at Tohoku University. He studied nonlinear physics and statistical physics at Nagoya University, and *Fractal* (Asakura Publishing), his thesis on fractal phenomena, is now considered essential reading on the topic. Later, he earned recognition through pioneering research into the new field of econophysics, linking the disciplines of economics, physics, and statistics through analysis of economic data using methodology developed in the physical sciences.

Mathematics genius Frank Nielsen, originally from France, investigates the geometric sciences of information, which is considered a framework for innovation in such fields as computer vision, medical imaging, and machine learning. Though the high-dimensional, noisy, largely heterogeneous, and non-Euclidean data with which he works require a great degree of specialist knowledge even by the standards of an organization such as Sony CSL, and as such seems beyond the reach of a broader audience, Nielsen continues to make an impact through the publication of numerous papers and other writings.

Each of these senior researchers is on a level with university professors or departmental heads. But, at Sony CSL, we have an array of young researchers with great potential as well as many who have already left their mark.

1.4.3 Blossoming of Youthful Talent

Alexis Andre is another Frenchman with a fluent command of Japanese. After earning his masters from a university in his home country, he received a PhD in computer science from the Tokyo Institute of Technology. At Sony CSL, the focus of his research is the leading edge of interactive technology, art, computer gaming, and design. At present, by combining existing and pioneering technologies, he is focused on "future forms of play" that go beyond mere toys to offer users genuine joy and enjoyment.

Shigeru Owada might just be the most unique researcher at Sony CSL. He received his PhD in information science and technology from the University of Tokyo and worked on interfaces for three-dimensional images and renderings of impossible objects through computer graphics. Since joining CSL, he has undertaken a variety of surprising side projects with titles such as "jello printer" and "toilet communication system." In recent projects such as Kadecot and Moe-Kaden, Owada continues to focus on the creation of fresh value in the field of smart houses and home appliance networks.

Statistics expert Takashi Isozaki is striving to revolutionize his field through the introduction of a thermodynamic approach. Several steps beyond the data correlation-based big data analysis that is currently finding popularity even among nonspecialists, the data science he is exploring is based on causal data analysis. His ultimate aim is to open up a new field by introducing the thermodynamic concepts of temperature and heat into statistics.

CSL Paris, the French branch of Sony CSL, is headed by François Pachet. A researcher at the leading edge of AI, Pachet is a musician and, through analysis of the structure of music, he has built a rich body of research combining these two themes. Other research at CSL Paris focuses on language learning and urban agriculture.

These are very brief outlines of the work of just some of the 30 or so researchers working at Sony CSL's twin labs in Tokyo and Paris.

And though the enterprise is relatively small in scale, the many talented staff with activities in diverse fields make for a multifaceted research facility.

1.5 SONY CSL-DERIVED TECHNOLOGY POWERING XPERIA™ SMARTPHONES

1.5.1 POBox Predictive Text Input, and One-Touch Devices Based on FEEL

What makes Sony CSL so extraordinary is that it doesn't just notch academic research achievements, it actualizes them in the marketplace through numerous new products, features, and lines of business. There are plenty of labs around the world doing revolutionary research, yet lacking the capability to commercialize their own achievements. But Sony CSL, with fewer than 40 personnel (even including TPO and other administrative support staff), packs a big punch when it comes to getting innovations to market.

For an example, look at Xperia smartphones, one of SONY's flagship product lines. It so happens that Xperia phones are jam packed with the fruits of CSL research (Figure 1.5).

They include a technology called POBox. When Toshiyuki Masui, now a professor at Keio University, was a Sony CSL researcher, he invented a predictive input method that enables long phrases to be entered with fewer keystrokes by matching partially inputted strings with past inputted phrases to present multiple candidates for auto-completion—a feature that is absolutely essential to using smartphones. It is based on a very simple and compact algorithm for pattern matching and becomes more effective the more you use it—to the point that many smartphone users would now be lost without it.

In 2000, Masui worked with SONY's mobile phone division to get the technology into a model called the au C406S. The technology went on to be used in Sony Ericsson phones and is still deployed in Sony Mobile Communications smartphone products, having undergone further development into POBox Pro, POBox Touch, and POBox Plus. As the last word in ease-of-use for text entry, not only is it one of the crucial capabilities of SONY smartphones, it has also been deployed in numerous other devices such as tablets and cameras (Figure 1.6).

Then there is one-touch operation technology, which supports, for example, functionality to automatically enable a Bluetooth connection and play music just by putting a smartphone and a speaker into physical

Birth of the Technology Promotion Office • 19

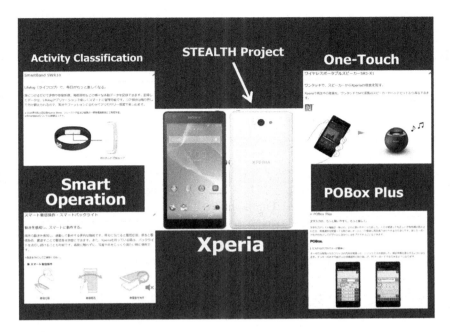

FIGURE 1.5
CSL technologies installed in the Xperia™. (Courtesy of Sony Mobile Communications, Inc. Tokyo, Japan.)

FIGURE 1.6
POBox predictive text input. (Courtesy of Sony Mobile Communications, Inc., Tokyo, Japan.)

contact (Figure 1.7). This is based on a 2001 research project by Jun Rekimoto called FEEL. Jun had the foresight to anticipate scenarios—so common today—in which multiple wireless networked devices are present in the same room and need a direct, intuitive means of connecting securely. So he came up with a concept that enables devices to initiate a connection over a close-range-only, authenticated near-field

FIGURE 1.7
One-touch connection. (Courtesy of Sony Corporation, Tokyo, Japan.)

communications protocol (such as FeliCa) and thereafter establish a high-bandwidth channel over an unauthenticated protocol such as Bluetooth or Wi-Fi.

First, via the near-field communication technology, the user is asked which devices he wants to connect with which, and the security keys and Internet protocol (IP) addresses are exchanged to set up the connections, which are then replicated to Wi-Fi or some other wide-field communication protocol. This has become the main means by which SONY smartphones connect with other SONY devices such as speakers, cameras, headphones, TVs, and so on.

Jun's approach has been adopted as a global standard and is now used by many non-SONY devices too. In Chapter 2, Section 2.2, I will address this technology in more depth.

1.5.2 Making Devices More Intuitive: Lifelog and Smart Operation Gesture Recognition

Xperia devices are now equipped with Lifelog functionality that uses vibration data from sensors such as accelerometers and gyros to perform activity classification—determining whether the user is running, walking, riding in a vehicle, and so on—at each point in time and to log it. At the 2014 Consumer Electronics Show, this functionality was announced in conjunction with SONY's SmartBand product. The technology is based on research done by Sony CSL's Brian Clarkson. When Brian started his research, the sensor bundle required filled a large backpack. Now SONY's SmartBand and smartphones pack all that into devices that fit in your palm or around your wrist (Figure 1.8).

The latest iteration of this functionality is called Smart Operation. This is an automatic gesture recognition technology that enables devices to

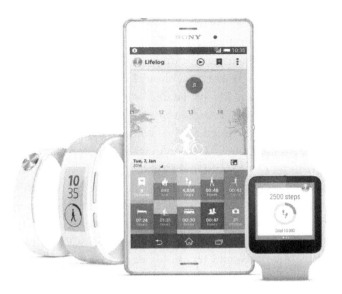

FIGURE 1.8
LifeLog. (Courtesy of Sony Mobile Communications, Inc., Tokyo, Japan.)

automatically execute the most appropriate behavior without needing any screen touches by the user: for example, intelligent screen orientation, accepting an incoming call when you raise your Xperia phone to your ear, rejecting a call when you give your phone a shake, muting the phone's sound when you put it down with the screen facing down, and so on. These behaviors are implemented by learning software that responds to accelerometer inputs resulting from the device's physical movements. The learning algorithms used are based on an engine originally developed by Francois Pachet of CSL Paris for analyzing music.

1.5.3 Sony CSL Is an Outlier among SONY's Many R&D Organizations

Sony CSL's research doesn't just flow into smartphones, of course. PCs, tablets, gaming systems, TVs, cameras, Walkmans, and many other SONY products have incorporated CSL innovations commercially. And CSL research has also led to technologies that drastically improve component quality by improving manufacturing processes, such as fixing bottlenecks in complementary metal–oxide–semiconductor fabrication and optimizing inspection equipment in factories.

Aside from CSL, SONY has many other researchers working to develop future products through innovations in domains such as next-generation semiconductors, battery materials, image compression, facial recognition, and many more. But even when tackling the latest and hottest tech fields, these research efforts are always geared toward creating more advanced versions of existing product lines, and they take their cues from SONY's business side, which is attuned to the "voice of the customer" and guides what research is undertaken based on market imperatives. This is research that begins with a desired outcome already defined.

In contrast, at Sony CSL, researchers are given the freedom to define their own topics—as long as they are shooting for the moon. In brief, CSL researchers set their own agenda. And that means that the scope of research topics they take on ranges way beyond SONY's existing businesses, venturing far into new technologies and fields that are not even on the radar of SONY business units.

Until not so long ago, the business side viewed this research unit that doesn't take requests and pursues far-horizon possibilities beyond even the most forward-thinking business plan as irrelevant. The result was a "valley of death" between SONY business units and CSL, a divergent agenda gap much greater than that with other SONY research units.

As depicted in Figure 1.9, the valley of death refers to the gap between R&D and commercialization, a concept frequently addressed in the

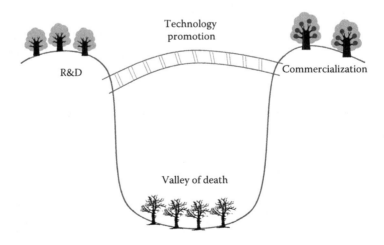

FIGURE 1.9
"Valley of death" between research and commercialization. (Courtesy of Sony Computer Science Laboratories, Inc., Tokyo, Japan.)

technology management literature. The classic scenario is that the research units make discoveries and write papers, while the business units are responsible for actually turning the discoveries into practical real-world products and bringing them to market. And yet, with many research successes, the business side never latches on to them, and they are buried without ever seeing the light of day. Yet the seeds of breakthrough products are found in these research achievements. In Chapter 2, I will talk about how we bridged the valley of death at Sony CSL.

1.6 MIRED IN OBSCURITY

1.6.1 Products Carrying CSL's Technology DNA, but Not CSL's Name

Ten years ago, Sony CSL won worldwide recognition in the scientific community as a center of cutting-edge research. Paradoxically, it suffered from a low degree of awareness within SONY.

CSL aimed to be the ideal research lab, bringing together trailblazing researchers from various fields. Unfortunately, it operated something like a strip mall composed of independent businesses sharing a roof. You had your cognitive neuroscience research shop in one storefront, your economics shop the next storefront over, and your biosciences shop on the other side. Only the director and deputy director of research had an overall picture of the research being done; therefore, the only people anyone from SONY could go to and ask about the research were two monumentally busy researcher/executives.

Meanwhile, over at the SONY mothership's business units, typical technology transfers from other SONY R&D shops were facilitated by personal connections among colleagues who were SONY lifers—people who had come up through the ranks together and remained drinking buddies. Sony CSL researchers, on the other hand, were mainly outside hires from academia or other companies, and they had few or no personal contacts in the business units. This kept business unit personnel largely unaware of Sony CSL's research.

Of course, Sony CSL wasn't just twiddling its thumbs. CSL's biennial open houses drew such throngs of SONY employees they practically had to be beaten away with sticks. And CSL also put on an annual research exhibition at SONY corporate that always piqued the interest of quite a few SONY engineers wanting to incorporate CSL innovations into products. In fact, some technologies did make it into products this way.

One example is the predictive text-entry technology POBox, which I discussed in Section 1.5. By 2004, it was in every mobile phone SONY sold in Japan. And yet almost no one at SONY was aware that the technology had come from CSL. People who had been directly involved knew, of course, and there was official recognition by SONY corporate headquarters, but Sony CSL's name was nowhere on any actual product featuring POBox, nor were any such products mentioned on Sony CSL's own website.

Even though the name POBox was trademarked, the trademark belonged to SONY Corporation. SONY Corporation was named in the trademark notices, and the name of Sony CSL was nowhere to be found. In fact, even though I worked at SONY and used SONY mobile phones myself, until I transferred to CSL, I myself had no idea that POBox technology came from CSL research.

In brief, not only did CSL research have very few points of contact through which research had the opportunity to advance toward commercialization, even those technologies that were lucky enough to do so did nothing to let the world, or the SONY rank-and-file, know that Sony CSL research was the source of these seeds of breakthrough, highly marketable innovations.

1.6.2 Changing Sony CSL's Research Style

Another important element was the shift in the nature of Sony CSL's research itself. In the 1990s, CSL was heavily into research on things such as Aperios, a novel object-oriented operating system (OS). Aperios had its research team transferred en masse from CSL to SONY corporate in order to turn it into a commercial product. The OS found its way into satellite TV receivers and was also used for AIBO robot technology (Figure 1.10).

That research team then moved on again to SCE, where they developed the OS for the PS. This kind of shuffling around of whole research teams to turn them into product R&D teams is standard operating procedure for corporate R&D that is common at SONY and many other big companies. But not all researchers are interested in being reassigned to product development; many would prefer to pursue a career in research. (This is especially true of the kind of researchers attracted to work at CSL, with its high ideals about what a research lab should be.) This is why POBox,

FIGURE 1.10
AIBO. (Courtesy of Sony Corporation, Tokyo, Japan.)

for example, was transferred just as a technology and not accompanied by the research personnel. This created a dilemma about how to achieve more effective technology transfer.

So Sony CSL's leadership decided to form a dedicated technology promotion team tasked with tackling the issue.

That was the state of things at Sony CSL when Tsukasa Yoshimura plucked me from treading water at SONY corporate to help set up a new organization within CSL.

1.7 TPO TAKES FLIGHT

1.7.1 Mission: Maximize the Fruits of CSL Research

On August 1, 2004, I transferred to Sony CSL to head up the newly established TPO, which had only Takahiro Sasaki and me as dedicated staff. Yuji Ayatsuka was splitting time between TPO and his user interface research, and Tsukasa was backing us up in an advisory capacity.

As a newly minted research sales force, we hurled ourselves into the work with a passion; but, with only two of us, we needed to dial in our limited manpower to focus on the most promising avenues.

Although Sony CSL's executives had clear expectations for TPO, across the organization there were varying interpretations, and researchers came to us hoping our new organization could do this or that for them. (An example: Can you come to this conference and video my presentation?)

And because the president/research director had no executive support staff of his own, unlike in a typical corporate research lab, we also received many requests from SONY corporate to attend various meetings. If we had accepted all those sorts of requests, Takahiro and I would have had our full bandwidth consumed right away and been prevented from achieving our core mission. So it was paramount that we define the work TPO would do right at the outset and make it clear to everyone that this mission would take priority. After much thought, we came up with the following TPO mission and action plan (Figure 1.11).

First, we defined TPO's mission as "optimizing technology transfer of CSL research achievements within SONY and maximizing their impact." We made a statement that we were going to find commercialization opportunities for CSL research at SONY and not go pitching outside the SONY group.

1.7.2 Three-Point Action Plan

Our action plan had these three points:

1. Take inventory of research.
2. Don't take a shotgun approach to selling; lock our sights on specific lines of business (gaming, mobile phones, PCs, etc.).
3. Maximize the commercial potential of research.

FIGURE 1.11
TPO mission statement when launched in 2004. (Courtesy of Sony Computer Science Laboratories, Inc., Tokyo, Japan.)

1.7.2.1 Point 1: Take Inventory of Sony CSL Research. What Did That Entail?

If we were going to be the sales force for CSL's research, we needed to understand the research. Often when pitching a technology, researchers accompany salespeople on the call, but that consumes researchers' valuable time. So ideally, the sales force should be able to explain the content of the research up to a certain depth on their own. That meant having individual meetings, with adequate time set aside, with all of the researchers on a regular basis. So Takahiro and I set out to understand the research going on at CSL.

1.7.2.2 What about Point 2: Lock Our Sights on the Most Promising Targets?

Sometimes—I've seen it happen many times myself—a business unit will try to entice a research unit with some kind of quid pro quo: We'll put your technology in our product, if you do such-and-such for us. Because researchers are passionate about seeing their innovations make it to market, a unit lacking its own research resources may want to harness a lab's manpower and use this force as an unpaid subcontractor.

Researchers tend to strive more than necessary to "talk up" the value of their research to business units that show an interest in the technology their study has yielded, whether or not those units actually have the resources to commercialize that technology. They see contributing to the business side as a feather in their cap. But these kinds of ad hoc, unpaid subcontractor relationships don't produce success stories. So even if it makes it harder to strike a deal, homing in on the right outlet, not just any outlet, for commercializing a technology is the path to real success.

1.7.2.3 And Finally Point 3: Maximize Impact

The example of POBox alone shows the importance of ensuring that a technology's Sony CSL origins be clearly exhibited and publicized. In the rare instance where a technology that a researcher created does get commercialized, it is a major source of frustration when no one realizes it. But it's not that business units are intentionally erasing Sony CSL's name, of course; it's simply that they don't think to credit CSL unless they are explicitly asked to do so. (This is demonstrated by the fact that when a technology from outside SONY is used, it is meticulously credited.)

In addition, the business units that adopt CSL technologies are usually development teams, while crediting on the web and on the product/packaging itself is handled by product planning and marketing people. Since it can take 2 or 3 years between the time a technology comes over and the launch of a product based on it, and considering the number of hoops the message has to get through in the organization, it is easy to see how the Sony CSL name gets lost in the shuffle.

So Sony CSL needs to clearly communicate how it wants to be credited, keep tabs throughout the saga of bringing a product to market, and speak up for its own interests in order to make sure the world knows that a technology comes from CSL. And the more awareness there is in other business units that Sony CSL technology made it to market, the more interested they will be in getting hold of CSL research for their own products. All of this is crucial work that TPO must do.

That was our philosophy out of the gate, anyway. But when the rubber met the road, the road turned out to be a bumpy, twisty one, and we had to shift gears many times in our mission and actions to make progress. We also learned many lessons from each project and developed an understanding of our work as a research sales force.

In the next chapter, I'll walk you through some cases that turned out well—and some cases that didn't.

2

Case Studies in Technology Transfer

2.1 VAIO POCKET: A PAINFUL LEARNING EXPERIENCE

2.1.1 Most Technology Transfers Go Nowhere

The number of projects that go nowhere are too numerous to count. To be blunt, 99% of our pitches of Sony Computer Science Lab (CSL) technology don't result in any concrete collaboration. And 99% of the few collaborations that do happen don't end up as products that reach the market. According to my score sheet, we are batting 0.001: Only one in 10,000 of the Technology Promotion Office's (TPO's) promotion pitches score a run for the CSL home team. So I have a rich and varied set of case studies on swing-and-miss projects.

Sometimes our approach is received with enthusiasm, but when we try to actually get the technology transferred, the business unit is too busy with other stuff to move things ahead around the bases. Even if the transfer actually happens, the business unit may end up stranding the technology on base, shelving it, and promising they will "get to it soon" but never making the further plays need to bring it home to score. Sometimes a technology gets all the way to the prototype stage, but one of the higher-ups quashes it before it gets to market. And even a product that has actually been announced publicly can end up as vaporware that never actually ships—called out at home plate, you might say. Technology transfer is like one of those marble mazes where there are holes at every turn waiting to drop the ball into oblivion on its way to the finish line. Despite that pitiful-looking 0.001 batting average, though, it beats batting 0.000: The only sure way to *never* get a hit is never to step up to the plate. So TPO keeps on trotting out there onto the field.

And even when we manage to navigate our marble past all those traps and roll it out into the marketplace for sale to customers, there are still pitfalls that can make CSL research drop out of sight without a trace.

2.1.2 Presense Technology and the VAIO Pocket

The SONY VAIO Pocket was a music player with a built-in hard disk drive (Figure 2.1). It went on sale in June 2004, not long before my move to Sony CSL. A user interface called G-sense that combined a touchscreen with a set of buttons was its claim to fame. And the G-sense interface was an implementation of research done by now-Deputy Director of Research Jun Rekimoto to create a technology called Presense, which did make it to market.

G-sense was a prominent feature of the design and the technological centerpiece of the device. Yet nowhere in the VAIO Pocket's user manual or on its product web page did the name "Sony CSL" or "Jun Rekimoto" appear. Moreover, none of the press coverage the Pocket received mentioned these origins. In short, there was absolutely no way anyone outside SONY would ever be able to know that G-sense was derived from Sony CSL research. To add insult to injury, the name change to G-sense from Presense was done without even informing Jun and meant that even someone who actively searched the Internet for information about "G-sense" wouldn't come across Jun's name since he had done the research on it under the previous product name.

Of course, the business unit had no ill intentions in doing this. As the technology was handed along from one department to another, the information about its Sony CSL origins simply got lost in the shuffle. And the name change was purely a marketing decision somewhere down the line. All understandable, but...

FIGURE 2.1
VAIO Pocket. (Courtesy of Sony Corporation, Tokyo, Japan.)

Nevertheless, we at TPO believed that it was crucially important for Sony CSL to receive credit for the technologies it produced. One reason that awareness within SONY of CSL was so low was that we never were formally credited even when our technology made it into shipping products, so there was no way for anyone other than people in the particular business unit involved to find out about the contribution CSL had made. In a way, it is like having to toil in obscurity for a pittance, and then, to make matters worse, have people call you a tightwad! As long as that state of affairs continued, researchers would be highly demotivated to take precious time from their research careers to cooperate in product development.

The launch of the VAIO Pocket came as a slap in the face to Jun and others who had been involved, and TPO felt stung. The product itself was thrown together in a very short period of time; it had only been a year since they had first asked Jun about incorporating his Presense technology. He must have put in a lot of effort to help them get the product out the door that fast. And *this* was his reward. Taking place around the same time that TPO was started up, this case served as an important learning experience.

In light of the VAIO Pocket fiasco, with subsequent technology transfers, we insisted up front on explicit crediting. For example, THE EYE OF JUDGMENT™ video game, which I mentioned in Chapter 1, credited Sony CSL, Jun Rekimoto, and Yuji Ayatsuka on the game's title screen and in its manual. In that instance, TPO had the business unit handle it appropriately.

2.2 THE DIFFICULTY OF TAKING TECHNICAL BREAKTHROUGHS TO MARKET

2.2.1 FEEL: A Landmark Idea

Now let me lay out the case of a technology called FEEL, in which TPO really had to run through brick walls to get the technology commercialized.

FEEL started as a research program for intuitively connecting devices wirelessly (Figure 2.2). Around 2000, Jun had already anticipated a future with myriad wireless devices everywhere around us in everyday life, and he launched the FEEL research project to create an easy way to network these devices together.

FIGURE 2.2
FEEL research project. (Courtesy of Sony Computer Science Laboratories, Inc., Tokyo, Japan.)

Since Wi-Fi signals have a range of tens of meters, if there are many Wi-Fi devices in a room, it's tricky to figure out which ones should wirelessly connect to which. For example, if you want to connect your mobile phone to some speakers via Wi-Fi, you usually have to switch both devices on, bring up a list of available devices to connect to on the phone, and pick the speakers from the list—a cumbersome undertaking. But when you are holding your phone with the speakers sitting right in front of you, there should be a more intuitive way to link them up.

FEEL solves this problem by combining an authenticated communications protocol with an unauthenticated one to implement easy connectivity. For example, when you want to connect the aforementioned phone and speakers, you just bring the two devices into physical contact. Using near field communication (NFC; an authenticated protocol operating over a range of a few centimeters, such as FeliCa), the devices swap Internet protocol (IP) addresses and security keys to establish a connection. Then that connection is replicated to Wi-Fi (a nonauthenticated protocol). All this means for the user is they just touch the two devices together, and they become connected via Wi-Fi (Figure 2.3).

In other words, the NFC protocol with a range of a few centimeters is used to assess the user's intentions (which devices he or she wants to connect), and then that connection is kicked upstairs to a wide-field communication

Case Studies in Technology Transfer • 33

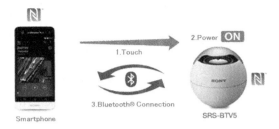

FIGURE 2.3
How FEEL works. (Courtesy of Sony Corporation, Tokyo, Japan.)

protocol (Wi-Fi) that can keep the devices connected without having to stay in physical contact. A key element of this technology is the concept of replicating a connection established on one wireless protocol to a different wireless protocol. That concept was actually invented by Mario Tokoro, CSL's founder, and is a key underlying patent of FEEL technology.

Mario's epoch-making idea touched off a project involving Sony CSL researchers, centered on Jun, along with designers and engineers from SONY Corporation. Prototypes were built of an IP phone, and monitors, PCs, and so on. Feel was even picked up by a project at SONY corporate for use in a keynote demo at the 2001 COMDEX trade show in Las Vegas. So, FEEL burst onto the scene in style.

But then the SONY project that had incorporated FEEL was scrapped due to company politics, leaving the technology itself in limbo. The canceled project at corporate seemed to mean the end of the road for FEEL. And that's when I came into the picture; having transferred to CSL's TPO, I took on the task of finding a new commercial outlet for FEEL.

2.2.2 FEEL's Lifeline: A Videoconferencing System

The FEEL project had foundered in a storm of company politics, but it found a precarious lifeline in a videoconferencing system that was the sole remaining FEEL-equipped product that continued to be developed. For TPO, which had just started up, FEEL was the perfect opportunity to prove our mettle, so we did everything we could to commercialize it. Finally, in March 2005, there was a press conference to announce the videoconferencing product to the outside world. We felt this was a prime opportunity to make the world aware of Sony CSL's technology.

However, the marketing people at the business unit told us that this was simply a presentation of a concept, and it wasn't going to be released as a product on the market. Considering that it was a whole new product concept, we couldn't argue. But then, on top of that, they also said that the Sony CSL name couldn't be used with the demo. Having poured so much effort into that technology transfer, I found that decision unfathomable and complained bitterly about it. The marketing people's response was that word had come down from the top brass at SONY that this was to be positioned only as a technology concept and would not be shipped, so they actually didn't want to draw much attention to it and didn't want to release any more information about it than necessary.

After protracted negotiations, we reached a compromise under which Sony CSL would at least be name-checked verbally in the concept product presentation. Even so, none of the media coverage of the prototype ended up including the Sony CSL name so, in fact, we gained zero visibility with the public. It was a painful defeat for TPO.

Later, further development of that prototype into a product was halted and the prospect of FEEL technology making it to market aboard a videoconferencing system evaporated. But that painful lesson spurred us on at TPO to codify clear conditions for technology transfers concerning how the products would be announced.

Incidentally, with regard to this incident, I had major confrontations with the person in charge of marketing at that business unit, and we had a prickly relationship. But when I looked back on it a few years down the road, I realized that we had both been fighting honorably from our corners, and I developed a kind of battlefield respect for that marketer. We went out for a drink a few times, and now I count him as a trusted friend both professionally and personally.

2.2.3 Adoption into the NFC Standard

We were feeling glum at TPO about how things had turned out when a new chapter opened—discussions about FEEL being incorporated into the NFC standard for which the FeliCa business unit was pushing. An NFC standard unifying a variety of methods for short-range wireless connectivity promised many potential applications in mass transit, transactions, and so on. Adding FEEL, the story went, would also turn NFC into an intuitive way of getting devices networked together. The previous project at SONY corporate that had adopted FEEL had in mind intuitive wireless connectivity that would work exclusively among SONY products. But if an official standard was established, a separate proprietary SONY approach would be out of the question; it would be in SONY's interest to agree with other companies on an interoperable standard. Jun and other members of the project team agreed and made a huge contribution by helping draw up documentation for the standard. People with experience in technology negotiations even became involved in hashing things out with other companies.

Having learned our lesson from the videoconferencing system fiasco, we first insisted that SONY corporate agree in writing to our conditions in the form of a Memorandum on Technology Release (MoTR). Whenever products using the resulting standard were announced to the public, Sony CSL would be credited by name. And, at that stage, our counterparts on the business side readily agreed.

These activities successfully led to the establishment of an international standard, and FEEL formally became part of NFC.

2.2.4 FEEL-Enabled Phones Hit the Market and the "MoTR" Saga

But even though FEEL was now part of an international standard, products actually implementing it failed to appear. At TPO, we continued to pitch it to all comers, and business units such as Sony Ericsson did express interest, but because FEEL would only become useful if featured in many different products—not just one—everyone was reluctant to be the first mover.

In the midst of this, out of the blue, a standard called CROSS YOU that implemented that the FEEL concept in Japanese mobile phones was announced (Figure 2.4). It was a communications standard for mobile phones to link up with each other, spearheaded by NTT DoCoMo (Japan's leading mobile phone operator), and numerous manufacturers announced devices supporting it. At Sony CSL, people who were familiar with FEEL's tortured history kept asking me, "Hey, isn't that FEEL?" So I scrambled to find out where CROSS YOU had popped up from. The answer turned out to be that the FeliCa business unit, the same one that had worked on

FIGURE 2.4
CROSS YOU: Two people can edit a photo at the same time. (Courtesy of Sony Corporation, Tokyo, Japan.)

establishing the NFC standard, had pitched CROSS YOU to mobile telecom carriers and succeeded in getting it rolled out.

But the people at FeliCa had not informed those of us over at CSL about any of this, and their conduct was clearly in violation of the signed agreement we had with them. As I dug further into the matter, it turned out that a young employee in their group, who hadn't been involved when the MoTR was settled, had relentlessly pitched the deal and made it happen. But the fact that he personally hadn't been party to the MoTR didn't change the fact that, as an employee of the FeliCa unit, he was bound by it and had violated it.

Ultimately, against ferocious resistance and tenacious foot-dragging by that individual, his boss and the whole FeliCa group, the SONY website and subsequent articles were amended to reflect the fact that Sony CSL was the source of the technology. Despite that employee's ignorance of the MoTR, its validity as an agreement between Sony CSL and the business unit was affirmed in the end, and action was taken to rectify the violation of terms.

The lesson we learned was that an MoTR was necessary, but not sufficient, to protect Sony CSL's interests. Over time, personnel changes at a business unit could result in people unaware of the MoTR picking up technology transferred from us at an earlier point in time. That meant we needed to proactively follow up and track the product development cycle from our end.

2.2.5 The Development of One-Touch

A few years after CROSS YOU debuted, the world of mobile phones underwent the seismic shift to the smartphone era. As we kept tabs on product development within SONY, we received word that research and development (R&D) on smartphones, speakers, cameras, and other devices equipped with unified FEEL connectivity was already at an advanced stage. Ultimately, these products were announced in the fall of 2012, but we found out 6 months before that, in early 2012.

At that time, one thing we learned was that since this effort was being carried forward by a cross-organizational team at SONY corporate, people who knew nothing about FEEL were running the show. This meant that we needed to bring them up to speed on the previous history of the research that went into FEEL technology.

Research on networking protocols such as FEEL is not embodied by a specific device or piece of software; rather, it represents a conceptual framework for linking diverse functions. Accordingly, from a product development person's point of view, since the product does not include actual code that came from the lab, there is a lack of awareness that the lab's technology is in fact being used.

Nevertheless, the approach the product development people are now implementing would not exist without cycles of prototype development and testing as well as incremental research—not to mention patents. The research side contribution cannot be overestimated.

In order to secure appropriate recognition of this reality, we had Jun go out and give a talk to the whole cross-organizational team. We positioned it as Jun sharing his insights into the technology from having shepherded it through the research stage; but the real purpose was to get the point across that CSL had done the research legwork on FEEL and to make sure that the people now pushing ahead with FEEL-enabled devices knew it.

And it worked. Even the FeliCa group that we had quarreled with over the CROSS YOU incident played nice, and everyone involved acknowledged that the new networking protocol they were taking to market was based on FEEL. Later, it was branded "One-Touch" (Figure 2.5). And coverage in the trade journal *Nikkei Electronics* mentioned its FEEL roots!

From 2000, when research commenced, to the release of a product with One-Touch networking in 2012, 12 years had elapsed (Figure 2.6). In 2000, Wi-Fi devices were in their infancy and Jun's research was far ahead of its time. But now that the world is awash with Wi-Fi, Bluetooth, and other wireless networking protocol signals, FEEL has become a crucial technology.

As high-flying researchers like Jun push the technology envelope ever outward, technology promoters like me are tasked with what might be considered the grunt work of finding the right timing and setting to bring these technologies to the marketplace. It takes a lot of swings and a lot of misses to get that one big hit. And then we doggedly work to ensure that, when technology *does* reach the marketplace, people know whose research made it all possible.

FIGURE 2.5
One-Touch connection. (Courtesy of Sony Corporation, Tokyo, Japan.)

38 • *Out of the Lab and on the Market*

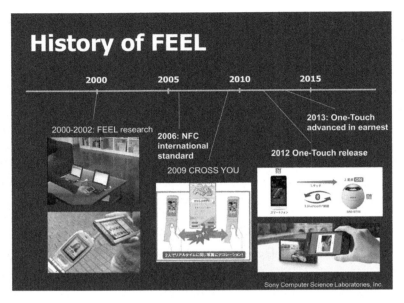

FIGURE 2.6
Path of FEEL to commercialization. (Courtesy of Sony Computer Science Laboratories, Inc., Tokyo, Japan.)

2.3 A CSL PARIS TECHNOLOGY'S UNEXPECTED ROUTE TO SUCCESS

2.3.1 EDS: Music Categorization Technology from Paris

Our next case study concerns a technology originating at CSL Paris. Sony CSL Paris is a small satellite laboratory hosting just five or so researchers. Headed by Francois Pachet, a leading figure in computer music research, CSL Paris has achieved research breakthroughs in language and music. In the early 2000s, Pachet invented a technology utilizing genetic algorithms for automated music analysis called Extractor Discovery System (EDS). Classified as a machine learning software, when fed in samples of, say, guitar or piano sounds, it automatically generates algorithms to distinguish between the sounds of the two instruments. In light of the ever-growing processing power of computers, TPO was convinced that this technology would prove extremely valuable (although we weren't clear on what specific applications it would have), so we made an enthusiastic push to promote a technology transfer.

EDS was featured at CSL's open house for SONY employees in Tokyo, and we pitched the technology to various SONY business units that seemed like a natural fit: the audio, PlayStation, and broadcasting equipment divisions, for example. But even though we drew some interest, it didn't lead to any discussions of concrete commercialization prospects. Suspecting that the language and distance barriers separating Paris from Tokyo might be hampering uptake, I demoed it far and wide—for the London division of PlayStation, an R&D group in Germany, a mobile phone unit in Sweden, and even the American unit of Sony Music and a software R&D group in Chicago. But everywhere we pitched it, we struck out. It's the nature of the beast at TPO to swing and miss a lot of times before smacking one on the sweet spot, but even so, this failure rankled.

When years of work had come up empty, I found myself moaning to Mario, who was still running CSL at that time, that EDS was a lost cause with too many intrinsic hurdles to pull off a technology transfer. But he encouraged me to keep plugging away and not to worry about how long it was taking, so I returned to the fray.

2.3.2 Building a Win–Win Relationship with SONY R&D

Meanwhile, I found out that Yoshiyuki Kobayashi, a "super-engineer" from SONY's Systems R&D Group, had been plugging away all on his own to expand the capabilities of EDS technology. Yoshiyuki and his boss at the time, Masaaki Hoshino, agreed with our initial assessment that the technology held a lot of promise, so Yoshiyuki went to Paris to spend a week working with Francois, was brought up to speed on the technology, and brought it back to Japan.

This development prompted those of us at TPO to give up on trying to achieve a direct technology transfer out of Paris, with all its difficulties, and to shift to a strategy of opening up wider prospects for EDS with Yoshiyuki's help. At the time, Yoshiyuki was engaged in a solo effort within SONY corporate's R&D group, so because he now had TPO pitching his work to various business units, it was a win–win situation.

I got a new demo video from Yoshiyuki and went around SONY trumpeting the progress he had made with EDS and introducing him to anyone who showed an interest. In this second phase, TPO's efforts at technology transfer did achieve results, making Yoshiyuki's work better and better known around the company. Moreover, Tamaki Kojima got on board and gave the software a more user-friendly package, dramatically increasing its usability and appeal.

Later EDS would end up contributing to the functionality of numerous products including mobile phones, tablet computers, Walkmans, and more. Not only did it show up in products, it also was used to develop visual inspection equipment for Blu-ray recorder production lines in Japan and Southeast Asia.

The most recent example of EDS technology is in the gesture recognition capability of Xperia phones' Smart Operation feature, mentioned on page 20. And more applications are showing up all the time.

The takeaway from this case is that a technology created at CSL Paris for analyzing music by a music researcher found far-flung and unthought-of applications in image analysis, vibration analysis, and manufacturing quality control. This surprising fact is not only a product of TPO believing in a researcher's work and promoting it but of a super engineer such as Yoshiyuki latching onto it and pushing it further to open up more opportunities within SONY.

What we learned is that even research that has taken flight from the research lab where it originally hatched and has winged its way to another organization through a technology transfer can encounter stiff headwinds that prevent it from soaring further onward and upward to the market, which is why persistent follow-up by TPO to guide it in to land in real-world applications is so essential. We call this follow-up "after-sale service." We keep in touch with the units and businesses where CSL technologies have been transferred and monitor their subsequent development. We ask the new stewards of CSL-derived technologies to make updated demo videos, and we maintain an ongoing working relationship in which our research sales force continues to feature those technologies in our road shows and to link up those who express an interest.

In order for an unusual "strip mall" research setup such as Sony CSL to turn its discoveries into real-world successes, connections with R&D organizations inside SONY are essential. EDS is a case study in success from following that principle.

2.4 INTERINDUSTRY COLLABORATIONS

2.4.1 Moe-Kaden: Giving Digital Appliances a Human Face

Of all Sony CSL's researchers, Shigeru Owada is, in my opinion, the most eccentric, off-the-wall individual—at a lab that prides itself on its many eccentric, off-the-wall individuals.

This is the man who created a three-dimensional printer that prints in edible gelatin (Figure 2.7) and a "toilet communication system" that enables someone in the bathroom to communicate with someone outside it. Let's just say no one other than Shigeru would have come up with these research topics!

At Sony CSL's open house (an event open to invited guests who are non-SONY employees) in 2010, Shigeru demoed his "Moe-Kaden" anthropomorphized appliance concept (Figure 2.8). Anthropomorphic anime-style characters of your refrigerator, TV, air conditioner, and other appliances appear on the Moe-Kaden app on a tablet; you touch the character to control the corresponding appliance. The appliance characters also interact with each other independently from user prompting. For example, in the demo, the avatar of a camera says to the avatar of a TV, "Show my pictures!" (even though the user is watching TV at the time) and a slide show of photos begins showing on the TV—to the consternation of the user.

FIGURE 2.7
Jello printer. (Courtesy of Sony Computer Science Laboratories, Inc., Tokyo, Japan.)

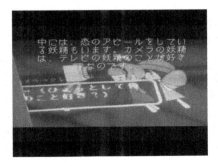

FIGURE 2.8
Moe-Kaden: assigning characters to household appliances. (Courtesy of Sony Computer Science Laboratories, Inc., Tokyo, Japan.)

One key official from Japan's smart electric grid initiative attending the demo observed enthusiastically, "This is what our SmartGrid project has been missing!" He introduced us to Daiwa House Industry Co., Ltd., a large Japanese homebuilder. Daiwa House had built a model smarthouse with all-digital appliances hooked up to a home server but was struggling for an application to harness that capability.

2.4.2 Barnstorming Negotiations with Daiwa House and within SONY

For TPO, this was our first big deal opportunity reaching outside SONY, and we worked feverishly to hammer out terms with Daiwa House while getting all our ducks in a row on the SONY side.

It turned out that the biggest obstacle wasn't cutting a deal with Daiwa House but dealing with people inside SONY who were vehemently opposed to Moe-Kaden on that grounds that it was in poor taste—that is to say, they felt that the "moe" style of the appliance anthropomorphization ("moe" being a Japanese slang term generally associated with cute anime-style cartoon characters) was not appropriate for SONY's corporate image. In other words, they just did not like "moe" stuff. It wasn't an issue of logic but of taste that made it that much harder to overcome.

Not only did we have to battle opposition inside SONY, we had to lay the groundwork by explaining things in advance to managers in the associated product divisions, make reports to the key executives, and so on. But all that explaining and convincing paid off in the end. In July 2011, at a Daiwa House facility in Suidobashi, Tokyo, Sony CSL and Daiwa House held a joint press conference about the collaborative venture and the story was picked up by the media in a big way (Figure 2.9).

As a result, Shigeru became well known as a pioneer in networked appliances. Many other companies bombarded us with requests for collaborative projects.

What TPO learned from working closely with Daiwa House to make the deal happen was that a whole new world of possibilities existed through collaboration with companies in totally different industries from SONY. For example, nobody knows more about building houses and punching holes through walls than Daiwa House; that's their bread and butter. But developing computer software is outside their wheelhouse. Sony CSL, on the other hand, has all kinds of expertise in creating software with novel functionality but doesn't know the first thing about homebuilding.

Case Studies in Technology Transfer • 43

FIGURE 2.9
Moe-Kaden: A joint project with Daiwa House. (Courtesy of Sony Computer Science Laboratories, Inc., Tokyo, Japan.)

By working together, we were able to envision a completely new type of product—a house with innovative digital features that neither of us could have created alone.

Daiwa House went on to initiate collaborative research with other CSL researchers, and they have continued to be an important open innovation partner for Sony CSL.

2.5 GOING TO MARKET WITH A PRODUCT TARGETING TEENAGE GIRLS

2.5.1 An App That Broke the Galapagos Barrier

12Pixels is an interface technology for creating pixel art on Japanese-style feature phones—phones that evolved in unique ways in the Japanese market.

Feature phones have an extremely limited interface that ordinarily would make it impossible to draw pictures on the screen in any intuitive way. But with 12Pixels, a very simple interface scheme allows the creation of pixel art. The typical feature phone has a keypad of 12 keys (the digits

0–9, *, and #) arranged in a three-by-four grid. The basis of 12Pixels is to have these keys correspond to 12Pixels on the phone's screen.

First, at the largest-scale layer, the phone's display is divided into 12 large pixels corresponding to the 12 keys of the phone keypad. By pushing the corresponding key on the keypad, you flip that pixel from black to white. Obviously a pixel this size would only allow an incredibly crude way of drawing, so 12Pixels lets you drill down to a smaller scale layer by using the enter key (the one in the center of the directional keys on the phone) and then use the directional keys to navigate to the area of the screen you wish to cover. Then you can flip smaller-scale pixels within that area using the 12 keypad keys, which allows more detailed two-dimensional drawings to be made (Figure 2.10).

This software was invented in 2007 by Sony CSL's Ivan Poupyrev and Karl Willis.

At the time, smartphones were not yet commonplace, and Japan was dominated by feature phones that eventually came to be dubbed "Galapagos" because the particularities of the Japanese domestic market

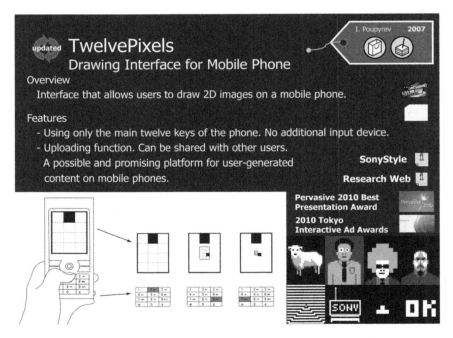

FIGURE 2.10
Outline of 12Pixels technology. (Courtesy of Sony Computer Science Laboratories, Inc., Tokyo, Japan.)

drove them to evolve in unique ways isolated from trends in the rest of the world. Before the advent of touch screens, a simple way of creating pixel art right on your phone was revolutionary. So we started pitching 12Pixels to various business units. But as soon as we explained that it allowed drawing, we'd be told that "drawing apps" for feature phones already existed. It was tough to convey a sense of how great 12Pixels was.

So, to make our case, we decided to actually distribute an app that would run on the feature phones of the time. This was something Sony CSL had never tried before, but the researchers were enthusiastic about it, so we developed the app for Japan's big three mobile carriers: DoCoMo, au, and Softbank. We validated it on a range of phone handsets. And we publicized its availability on the Sony CSL website. Since the Sony CSL website doesn't exactly reach a wide sector of the public, we worked with Sony marketing to have a beta version of the still-under-development app added to the *Taiken Kukan* (Experimental Corner) area of the main SONY website.

At the same time, we created a gallery where users could upload art created using the app. Unfortunately, the initial art that was uploaded seemed kind of stodgy. That was because the *Taiken Kukan* area tended to draw only the most dedicated fans of SONY products, a customer segment with little or no overlap with the teenage girls we expected to be the most active users of the app.

2.5.2 Proving the Power of a Public Beta

Meanwhile, I happened to meet someone who worked in the automated photo booth business (known as *purikura* in Japan) and, when I told him about our app, he was enthusiastic and agreed to promote it in an e-mail newsletter that his company put out (in exchange for which I agreed to connect him with somebody from Sony Music). That e-mail newsletter was the spark that enabled the app to catch fire among teenage girls, and a totally different type of art started appearing in our online gallery.

Examples included pixelized "boy + girl"-style love notes, the logos of chain restaurants such as McDonald's and KFC, portraits of pop stars, you name it. And the actual number of submissions went through the roof; more than 30,000 pieces of user-generated pixel art poured in (Figure 2.11). Since users could also store art on their own phones, the app must have been used to create at least 10 times that number of artworks. The reception was far more enthusiastic than we had imagined.

46 • *Out of the Lab and on the Market*

FIGURE 2.11
Gallery of works made using 12Pixels. (Courtesy of Sony Computer Science Laboratories, Inc., Tokyo, Japan.)

Based on those results, we took the app once again to Sony Ericsson in Sweden, and they told us, "This looks promising! We want to roll this out on our handsets globally!" (Not long before they had told us, "This is a dud!")

So 12Pixels was included in the Cedar handset that was sold in more than 40 countries across Europe, the Americas, the Middle East, India, China, and beyond. It was enjoyed by countless users (Figure 2.12).

Unfortunately, when Sony Ericsson subsequently switched over to exclusively producing smartphones, 12Pixels became defunct; but at one point,

Case Studies in Technology Transfer • 47

FIGURE 2.12
Global feature phone Cedar™ with 12Pixels installed. (Courtesy of Sony Mobile Communications, Inc./Sony Computer Science Laboratories, Inc., Tokyo, Japan.)

it reached a very wide market and I'd say that qualifies it as a success story. Moreover, the researcher was able to produce a scientific paper from it by analyzing pictures generated using the app and so on that won the "Best Presentation Award" at a conference called Pervasive2010. So it was a success on the academic side as well.

What was the lesson of 12Pixels? We would never have had the same outcome if we had just developed the technology through research at the lab and only tested it with people in our immediate circle. By throwing it open to the public, and soliciting and publicizing user-generated content, we were able to generate momentum that propelled it into a shipping product. We realized the potential power that a public beta could have for a research lab such as Sony CSL, and we created an "Online Experiments" section of the Sony CSL website that became a platform for subsequent research in various forms.

Now we are living in a world where releasing beta versions to the public as early as possible to gauge user response is a commonplace strategy, but 12Pixels was an example of how that strategy could work for a blue sky research lab and not just for startups.

3

Next-Level Challenges for the Technology Promoter

Typically, a corporate research lab's contribution to a business is seen as putting new technologies on a path to becoming part of a shipping product. But my experience to date in technology promotion has taught me that there are actually a number of different patterns for a research lab to follow to contribute to its corporate parent's business. Here are some examples of contributions that took forms other than supplying fresh intellectual property (IP) grist to the mill of new product development.

3.1 ECONOPHYSICS OPTIMIZES SEMICONDUCTOR PRODUCTION

3.1.1 It's Called Econophysics: Now Where Can We Use It?

Broadly speaking, at Sony Computer Science Lab (CSL), we have two kinds of research: systems research and theoretical research. CyberCode and other engineering research projects I've discussed so far fall under systems research, but CSL also houses researchers doing theoretical work in fields such as statistics, biology, brain science, and econophysics.

Ten years ago, when Sony's Technology Promotion Office (TPO) was launched, I was talking to a trusted senior colleague who told me, "You can usually imagine commercial possibilities for systems research. But theoretical research? Very hard to imagine any commercial potential."

I myself had the sense that it wouldn't be easy to commercialize discoveries made in those more abstract disciplines, but I wasn't willing to give up on them. So I buckled down to study up on what theoretical research was happening in our shop and came up with various ways to apply those researchers' discoveries to SONY's business operations.

One pillar of Sony CSL's theoretical research is called econophysics, and it's the bailiwick of Senior Researcher Hideki Takayasu (Figure 3.1). He applies advanced mathematical methods from physics and statistics to economic phenomena—such as exchange rate movements and transaction networks—in order to identify underlying mechanisms that parallel physical systems. For example, when we look at exchange rate movements, we find a large number of currency dealers quickly acting on all kinds of information to make buy-and-sell decisions; their collective decisions establish market prices among traded currencies. Those overall currency movements include features that echo phenomena in physics. Econophysics is a new way to gain insights into those features of economic phenomena.

This research approach is based on identifying rule-governed behaviors by using various methods of crunching enormous volumes of data. Of course, Hideki has his own sources for data from the banking and financial worlds, but his constant refrain is, "Give me more data!" Taking him at his word, I eagerly planned to find new sources of data at SONY on which he could put his methods to work.

My simplistic initial assumption was that since Hideki worked mainly with financial institutions, I should get him together with SONY's finance people. But when I approached them, it turned out that the only currency exchange activity they had was executing a basic array of exchange rate contracts necessary to support SONY's global operations; they were not conducting currency transactions on anything like the scale of banks or other financial institutions where the application of Hideki's methods would make sense.

FIGURE 3.1
"The Discovery of Econophysics" by Hideki Takayasu. (Courtesy of Kobunsha, Tokyo, Japan.)

So I changed tack and started looking around for any operations at SONY that involved handling huge data volumes, financial or otherwise. Some ideas that came to mind were sales data or materials procurement data. Then one day at an open house event, I happened to read a comment by someone who mentioned offhandedly, "Semiconductor manufacturing plants generate a vast amount of data that are hard to wrangle." So I put semiconductor manufacturing on my target list.

3.1.2 Forging a Unique Partnership with the Semiconductor Business Group

Later, I happened to meet a manager in SONY's Semiconductor Business Group (SBG) named Toyohiro Tsunakawa. As it turned out, that was destiny. Toyohiro had originally talked to Tsukasa Yoshimura about a certain chip, and Tsukasa then introduced us. At that point, Hideki wasn't in the discussion at all, but a few days after Toyohiro and I had connected, he and I hatched a plan to apply Hideki's research to semiconductor manufacturing. Toyohiro himself, with his expertise in the semiconductor fabrication process, got a gleam in his eye when he heard about Hideki's research.

So Toyohiro and I started to go around chatting up various people in management at Sony CSL and the SBG and urging them to apply Hideki's methods to analyzing data from the semiconductor fabrication process. But Sony CSL, the pure research lab, made a strange bedfellow for a semiconductor fab, completely focused on operational nitty-gritty. The cultural differences made it tough to get management buy-in on both sides.

I can still vividly remember the meeting I had with Mario Tokoro, who was CSL's director at the time, Hiroaki Kitano, who was then deputy director, and Hideki Takayasu himself, who was a senior researcher. Hideki and I had really honed our pitch. We explained why tackling a project together with the SBG was worthwhile. But Mario and Hiroaki were dead set against it. They grilled us with questions: "Considering we have no expertise in semiconductors here, isn't it a stretch that we can deliver results for them?" "If we go into this half-cocked, it's going to backfire." But Hideki and I dug in our heels and insisted we wanted to go forward with it. In the end, we got the go-ahead... along with the admonition, "If you're going to do it, just do it! But do it right and get results!"

Having obtained the go-sign from management on both sides, Toyohiro took the reins of the project and arranged a weekly videoconference between Hideki and the semiconductor plant in Kyushu.

A semiconductor fab is a goldmine of data. Better analysis of that data leads directly to improved production yields, and improvement in yields goes straight to the factory's bottom line. Hideki brought the most advanced statistical modeling techniques from econophysics to bear, working with the SBG and a team of engineers at the factory to analyze the data and improve yields. It didn't hurt that Hideki was originally a physics professor. His scholarly determination to get to the bottom of bad lots and find the root cause through rigorous analysis of observation data was a vital element.

Thanks to the efforts of Hideki and Toyohiro, as well as the team on-site at the Kyushu factory, within a year of the project's start in 2005, major improvements in yield had been achieved at the semiconductor fab. At that point, even people who had been skeptical previously acknowledged that the effort had paid off.

3.1.3 A Top-Shelf Example of Bottom-Line Gains

Nine years later, the project continues to yield benefits; the weekly videoconferences are still being held, and recently algorithms that came out of the project have been incorporated into the production management system on a line in Kumamoto. This is a concrete example of theoretical research delivering business results.

This example showcases the possibilities for analytics to drive process improvement and directly contribute to profits. In fact, this project has been lauded as one of Sony CSL's clearest winners in terms of profit contribution.

Within a large organization, the field of view individual employees have of what is going on at the company can be surprisingly narrow. Within the same team or a single research laboratory, there are events that bring about chance meetings among colleagues. And collaboration can arise naturally among organizations that are closely linked. On the other hand, colleagues who are within disparate units such as research and production have virtually no natural opportunities to mingle in the normal course of their work. Nevertheless, the most appropriate outlet for a research discovery may be lurking in a completely different part of a sprawling company.

From the econophysics semiconductor case, we at TPO learned the critically important lesson that a sales organization representing a research unit must always maintain a wide network of contacts to facilitate connecting normally far-flung parts of the organization, such as researchers and production engineers. This demands daily effort to cultivate sustained personal connections that overcome initial dismissal ("Your part of the company has nothing to do with ours!") and inspire openness ("Maybe we *can* get together on this").

3.2 A NEW OUTLET FOR RESEARCH DISCOVERIES: SCIENCE CONTENT FOR ENTERTAINMENT MEDIA

3.2.1 Ken Mogi's "Aha! Experience" Draws Attention and Sega Wants In

Next, I want to tell you about the Aha! Experience Project of Ken Mogi. Ken is certainly the best-known researcher associated with Sony CSL among the general public in Japan. Along with his academic research in the field of brain science, he makes frequent appearances on TV and is a prolific author. His high profile in the public eye makes some people at SONY wonder how he finds time to actually do research when he is so busy being a celebrity scientist. But from firsthand observation, I've concluded that Ken has the capacity to do 10 times as much work as an average person—which, by my math, gives him enough capacity to be three times better than average as a researcher, three times better than average as an author, and three times better than average as a TV personality, all at the same time, with room to spare. He also has a remarkable gift for complete focus on whichever thing he is doing at the moment. That's why he's a genius, right?

A buzzword that Ken brought into currency in Japan is the "Aha! Experience." A frequent feature on Japanese TV shows, the Aha! Experience consists of a scene that changes only a tiny bit at a time so that even though the starting and ending appearances are very different, you don't notice the change as it is happening. This setup was created by brain scientists as a means for artificially inducing an "Aha!" or "Eureka!" experience in the mind, but Ken brought it to TV and turned it into a pop culture sensation overnight.

One example of an "Aha!" moment is how the idea of gravity came to Newton when he saw an apple fall. But you cannot predict or control when Aha! moments will take place. However, by introducing a series of localized changes in a 15-second image, and having the participant concentrate on detecting where the change occurred, it is possible to induce an Aha! moment. This is known as an "Aha! Experience." In Japan, the entertainment side of the Aha! Experience is now far better known than the scientific side.

In 2005, after Ken had demonstrated the Aha! Experience on a TV show, an offer came in from a game development team at what was then known as the Sega Corporation to create a game for the PlayStation Portable (PSP). I was firmly opposed to letting Sega develop a game with Ken's name on it because it seemed only right that something like that should come out of SONY's own game development unit. But Sega had a concrete plan to get the game out, and Ken himself liked Sega's proposal. So I consulted people who really understood the video game industry and was advised that going with Sega would be best since they already had the game planned out. I was convinced and decided to go ahead.

3.2.2 Tapping Sony Music Artists to Do a Deal with Sega

But we were sailing uncharted waters: A SONY research lab collaborating with an outside company to make a video game was unheard-of, so we anticipated numerous obstacles. Moreover, as a research unit, we had absolutely none of the capabilities in-house for handling rights to people's likenesses, managing revenue streams, and so on. But Mario pointed out that the SONY Group had companies inside it doing talent management, and that we might be able to turn to them for help.

At that point, the Sega side expected that they would handle the entire architecture of the game, such as the graphics, while Ken would supervise the content. In other words, it wasn't a technology licensing agreement that they wanted, only Ken's input on content and consent to use his likeness—matters that would ordinarily be handled by a talent agency. So I consulted some people at Sony Music that I had worked with on FourthVIEW and they pointed me to a department within a SONY Group company called Sony Music Artists (SMA) that has a division representing nonmusician talent.

I headed straight over to SMA and brought the person I had an introduction to there up to speed, requesting that he manage affairs between Sega and Sony CSL exclusively pertaining to the Aha! Experience. Thankfully, he said yes. From that point on, all the stuff about the Aha! Experience that involved entities outside SONY was handled by SMA. Their first task was to broker a deal between the TV network whose show Ken had appeared on and Sega; they easily put together a win–win deal that all sides were happy with, for which we were very grateful. At TPO, we were completely clueless about show business and would hardly have known where to begin!

With the tie-up with Sega settled, there remained a need to explain to numerous departments inside SONY how this totally unprecedented deal was going to work. Human Resources, Legal, Finance, Marketing, Brand Strategy, Technology Strategy, and so on and so on... I must have given my presentation about the Aha! Experience video game business plan 20 times or so. But although some departments created hoops for us to jump through, others were unexpectedly cooperative. In the end, I got all the key people to sign off on it.

Meanwhile, I told Sega that if we couldn't get our internal approvals, we might write off the whole project, but waiting for approval to start developing the game would fatally delay it, so they went ahead with development anyway, acknowledging the risk. And, in fact, the okay did not come on our side until a month before the game was finished—and 3 months before it went on sale. Too late, in other words, to back out! My counterparts at Sega must have been sweating bullets.

On June 22, 2006, the PSP game "Sony Computer Science Laboratories' Dr. Ken Mogi Presents: Aha! Experience Games to Entertain Your Brain!" was released (Figure 3.2). Although it required a compressed development schedule, the release was timely enough to capitalize on public interest in Aha! Experience while it was still riding high. The game was well received and a sequel was made too.

FIGURE 3.2
PSP® game "Sony Computer Science Laboratories' Dr. Ken Mogi Presents: Aha! Experience Games to Entertain Your Brain!", ©SEGA. (Courtesy of SEGA, Tokyo, Japan.)

3.2.3 An Aha! Experience Ecosystem with an Eight-Digit Dollar Value

That could easily have been the end of the story. But it wasn't. The video game proved to be the starting point for a cascade of bigger and bigger opportunities, much as in the Japanese fable "The Straw Millionaire" in which a poor man barters his way from a piece of straw to a fortune. First of all, at the same time the video game was released, a museum in Tokyo's Odaiba waterfront district called Sony ExploraScience opened a temporary exhibit called Aha! Experience Square. It featured the Aha! Experience PSP game, Aha! Experience content, and explanations of how the games work based on brain science. The exhibit was a hit and the manager of the Sony Building in Ginza asked whether they could host it after its run at ExploraScience. So the Aha! Experience event moved on to the Sony Building.

Then Sega came to us because they also wanted to make a mobile game. Of course, we brought in Sony Ericsson from the SONY Group to collaborate on that. The app was available for download on a Sega site called Sonic Cafe and on Sony Ericsson's PlayNow site. That too was a hit, with one result being pre-installation of the game on the Sony Ericsson W51S handset for Japanese mobile carrier au, a move that generated tons of buzz (Figure 3.3).

By this time, proposals were pouring in from outside companies. In spring 2007, Otsuka Pharmaceutical Co., Ltd.'s ad campaign promoting its Calorie Mate line of nutrition bars to students cramming for entrance exams included Aha! pictures in the packaging. That paved the way for the display of a giant Aha! picture in Ginza to promote a new song by pop star Ken Hirai... Aha! pictures in Sourcenext's New Year's greeting card software... an Aha! movie trailer for the film "Night at the Museum"... an Aha! movie promo for a dental gum product by Sunstar Inc... and more. Dealing with this deluge became part of TPO's duties (Figure 3.4).

FIGURE 3.3
Aha! Experience (pre-installed on W51S mobile phone). (Courtesy of Sony Mobile Communications, Inc., Tokyo, Japan.)

Next-Level Challenges for the Technology Promoter • 57

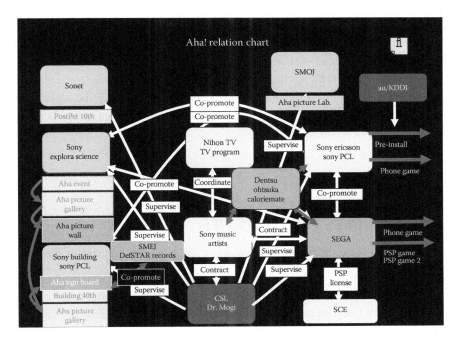

FIGURE 3.4
Aha! Experience: Contracts and other business relationships. (Courtesy of Sony Computer Science Laboratories, Inc., Tokyo, Japan.)

In all these deals, SMA acted as the intermediary while TPO supported production and Ken supervised. TPO and SMA made a big effort to minimize any burden on Ken aside from supervising the content. But it's safe to say none of this would have been possible under the auspices of a conventional Japanese research lab.

Meanwhile, on the business side, Sony Computer Entertainment (SCE), Sony Ericsson, and other SONY Group companies, plus SMA who was helping us do these deals, Sony PCL, and outside partners such as Sega and Otsuka developed a web of interlocking relationships that can only be described as an Aha! ecosystem with an economic impact in the tens of millions of dollars. The social impact was also substantial. These achievements obviously rested in large part on Ken Mogi's personal fame, but we certainly helped turn that name recognition opportunity into real revenue streams.

The significance of these projects to TPO was to validate "science content" as a new type of commercially significant research output. Aha! Experience content depends on a certain degree of know-how to create it but not on any complicated technology. But the synergy between Ken's persona and explanations and the content itself transforms it into a new and much more appealing genre of "science content." It was the proposal from Sega that first opened our eyes to this and provided the stepping stone to all the other developments that fanned out from it. The Aha! moment inspired in us by the Aha! Experience business development case: If you want to commercialize research discoveries, you can't be a prisoner of conventional thinking. You have to seek out new avenues to market that are off the beaten track.

3.3 SONY CSL'S FIRST SPIN-OFF

3.3.1 The Place Project: A New Kind of Location-Sensing Service

Up to this point, I've been covering examples of research discoveries that were commercialized by transferring them somewhere. That approach is predicated on there being someplace suitable to transfer them to. But with totally new fields, or new technologies that span multiple businesses, the right home within the organization may not exist. So now I'll turn to an example of a spin-off, in which a whole new company was birthed out of CSL.

Sony CSL brings together people from quite a range of backgrounds, and researcher Taka Sueyoshi is one interesting example. He was not a dyed-in-the-wool researcher from the get-go; he previously worked in management at a research and development (R&D) unit within SONY corporate. But he boldly transferred out of that post in 2005 with a mission to launch a new business out of Sony CSL. After scoping out the different research going on around the shop, Taka teamed up with two researchers already at CSL, Atsushi Shionozaki and Jun Rekimoto, to start the Place project.

The Place project arose from a technology called PlaceEngine, which enables a device to use Wi-Fi signals to determine its physical location.

In today's world, urban areas are awash in Wi-Fi signals. There are public wireless local area network hotspots from operators such as SoftBank and Wi2, along with access points belonging to companies, stores, or private homes (Figure 3.5). Nowadays, when you pull up a menu on your device to select a wireless connection, you will usually see a long list of access points within signal range of your current location. Of course, since each access point is secured with its own password, you can't just hop onto any one you want to establish Internet access. But the existence and the strength of their signals at a given location is information accessible to everyone.

FIGURE 3.5
Wi-Fi distribution in Tokyo. (Courtesy of Koozyt, Inc., Tokyo, Japan.)

Wi-Fi signals are quite short-ranged—in an urban setting, probably less than 100 meters. That means the set of access points that can be detected will vary from place to place around the city, which makes that information like a fingerprint uniquely identifying a location. Turning that around, with information on the location of access points and the set of access points that your device can currently catch a signal from, its physical location can be determined.

But there is no existing master database of Wi-Fi access points and their locations. Wi-Fi access points can be set up by anyone, anywhere; they may be turned on only part of the time. And, when a company relocates its premises, its Wi-Fi access points go with it. All of which is to say that the Wi-Fi access point landscape is ever-changing.

That is where PlaceEngine technology steps in, with a crowd-sourced database that anyone can update—like Wikipedia. First of all, this allows known access points detectable from a given spot to be explicitly registered as learned data. For example, when querying the database, if an unknown access point is also detected near a known access point, the database can be automatically updated with the previously unknown one. In other words, as the users use the system, the database is constantly updated (Figure 3.6).

We took this technology to a SONY product group and there was interest in using it. But that product group couldn't build a business based on it because managing a widely used location-sensing service was outside the scope of their product category and beyond their capabilities. In addition, there was doubt about whether a project under a research lab would have long-term continuity. In light of this reception, and in line with the inclinations of Taka and of Sony CSL's executives, we decided that Sony CSL should spin off a startup company offering a unique location-sensing service.

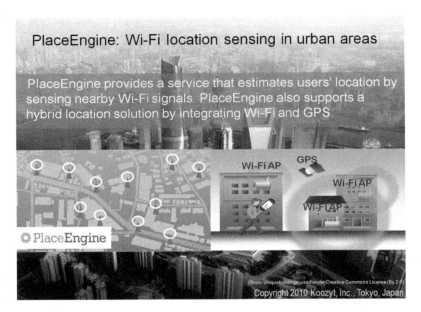

FIGURE 3.6
PlaceEngine. (From Koozyt, Inc., Tokyo, Japan. With permission.)

3.3.2 Creating a Buyer

That is when things started to get tricky. It was the first time we'd done a spin-off, so there was no blueprint. Within a large organization, things that have never been done before face all kinds of resistance. So it was not easy to push this spin-off forward. We had to grope our way forward via trial and error. That's how we stumbled on an appropriate company within the SONY Group: So-net, an Internet services provider with its own listing on the stock market and an independent business model. So-net would provide the capital and Sony CSL would provide the technology and personnel to establish the spin-off company.

But the final hurdle was how to get approval from SONY's senior management, who held final sway over whether the new company could be formed. I still remember the late-night sessions where Mario, Taka, and I hammered out a strategy. In the end, exploiting Taka's background of having worked under a certain executive at SONY corporate, we decided the best option was to have Taka simply go in cold and bring it up. So the next day, Taka dropped by headquarters to see his old boss, chatted amiably with him about old times, and then asked if he would go to bat for getting the spin-off approved. We all breathed a sigh of relief when Taka brought back good news.

FIGURE 3.7
Koozyt, Inc. founders: Rekimoto (left), Sueyoshi (middle), Shionozaki (right). (Courtesy of Koozyt, Inc., Tokyo, Japan.)

FIGURE 3.8
The logo of Koozyt, Inc. (Courtesy of Koozyt, Inc., Tokyo, Japan.)

Having surmounted numerous obstacles, in July 2007, Koozyt, Inc., was founded. Taka Sueyoshi was named the chief executive officer and Atsushi Shionozaki the chief technology officer. Both went to work for the new company, while Jun Rekimoto was placed on the board and made a technology advisor. Koozyt, Inc., is an integrated technology solutions provider that supplies software running on various devices in accordance with customer needs, and it also runs the servers needed for location sensing (Figures 3.7 and 3.8).

Koozyt has grown into an influential company in supplying the marketing industry with location-sensing and augmented reality technologies that work inside buildings, notably large retailers and museums, as well as face and facial expression recognition technology. Using its technologies, Koozyt provides various services on smartphones including advertising, promotions, and navigation.

3.3.3 Forging a Flagship Product for the Tokyo National Museum

Yoko Honjo of TPO played a big part in the success of the Koozyt spin-off. She joined TPO just after Koozyt was founded and was a prime mover in getting it off the ground. Yoko was involved in various initiatives to create early applications of PlaceEngine. The flagship was a project for the Gallery of Horyuji Treasures at Tokyo National Museum. It is still in use today—a testimony to her tireless efforts.

Yoko believed from the outset that a flagship deployment was needed to demonstrate the potential of in-building location sensing. She locked in on the Tokyo National Museum, the crown jewel of Japanese museums, with the idea of providing a navigation guide. But nothing is ever simple, and it took prolonged negotiations with the museum and other parties, as well as plenty of patience.

The Tokyo National Museum in Ueno, Tokyo, has vast collections, and some objects are rotated in and out of display as frequently as once a month. So the first trial iteration, lasting 3 weeks, was limited in scope to the Gallery of Horyuji Treasures, using terminals that museum guests could borrow free of charge during their tour. This real-world demonstration used PlaceEngine technology to automatically cue up videos about important items on display based on the user's location (Figure 3.9).

From there, as recognition of the navigation guide system's effectiveness spread, its coverage was expanded to the entire museum. Multimedia content was added to create digital "hands-on" interaction with artifacts, such as showing what kind of sound an instrument generates or the

FIGURE 3.9
Navigation system used at Tokyo National Museum. (Courtesy of Tokyo National Museum, Japan.)

process by which a piece of art was made. Now branded as "Tohaku-Navi" ("Tohaku" being an abbreviation of the museum's name in Japanese), it is the museum's official free digital guide.

Yoko served as more than just a matchmaker on this project. She was in a wide-ranging producer role overseeing everything from content planning and creation to service rollout. With her exceptional combination of expertise in the arts and technology, she was the perfect person to drive this project. And she has gone on to work on other initiatives at CSL where art and technology intersect.

This case study in technology promotion went beyond simply pitching research discoveries to business units. It required extremely challenging work such as creating a business plan for a startup company, lobbying for approval from SONY's top brass, finding a source of investment capital inside SONY, assembling a flagship product for launch, and much more. Only the relentless determination of CSL's executives, and of Taka and Atsushi themselves, made it possible. The Koozyt case stands as an example of the importance to a research lab of embracing business incubation. It's essential for a research operation that is constantly creating new things to be willing to create a new entity, in the event there is no place in the organization for a discovery to go to realize its commercial potential.

3.4 TOWARD A NEW ELECTRIC POWER INDUSTRY

3.4.1 Creating a Next-Generation Electrical Infrastructure

A CSL initiative that aims to pioneer a truly huge new industry is the Open Energy Systems (OES) Project. Sony CSL Founder Mario Tokoro personally runs the project, with an ongoing mission of changing the world. But when OES first got started 10 years ago, it was just one line of research from one CSL researcher whose progress TPO was tracking, like all the rest. This project constitutes our final case study.

At the time of this writing, OES is the biggest ongoing project at Sony CSL. The infrastructure that currently provides electric power to homes and businesses in Japan, like that of other industrialized countries, is the opposite of an "open energy system." It is closed, top-down, and centralized, consisting of giant power plants generating massive amounts of electricity that is delivered via a distribution grid to vast numbers of remote consumers. In contrast, OES follows a principle of generating power close to where it is consumed, using a bottom-up, open, and distributed electric power grid. Instead of a unidirectional system where power flows strictly one way, from upstream producers to downstream consumers, OES envisions many power users possessing small-scale production capacity in the form of solar panels, wind turbines, and so on—all interconnected by a bidirectional smart power grid that enables them to exchange electricity—an "Internet of electricity," if you will.

Let me give you a comparison: The legacy electric power system we have is like old-fashioned TV broadcasting. The studio makes shows and puts them on the air for all the viewers to tune in simultaneously. But now, with YouTube and other new digital media, it's become possible for individuals who were once passive viewers to make their own content and actively make it available for other viewers to watch. It's not that conventional TV broadcasting has vanished, but the rise of new media has given people more choices of what to watch. The underlying technology that made that possible was the Internet. At Sony CSL, we intend to pursue for electricity the path of evolution already taken by video content—from top-down to bottom-up.

The rising share of power supplied by renewable energy sources such as solar PV panels is an essential element of the new system we envision. The current centralized power grid is designed to deliver electricity from power plants at the hub to users at the edge. This means that there is limited capacity for accepting power generated at the edge back onto the grid (reverse current flow). As a result, in many places, net metering schemes have been halted or subjected to limits. This means that, for the share of renewable energy to keep rising, a new electric grid will be needed that is like the OES being pursued at Sony CSL.

3.4.2 A Large-Scale Project Involving Academic and Corporate Partners

This research is not at all ivory-tower, castles-in-the-sky stuff. As a matter of fact, at the Okinawa Institute of Science and Technology Graduate University (OIST), there is already a test-bed cluster of 19 real houses, with real people living in them, connected by a microgrid. The system has been running smoothly since late 2014.

OIST hosted symposia on the topic of OES in 2014 and 2015 that drew scholars from leading research institutions in many countries, representatives of energy industry corporate players, and other experts and policymakers. The key aim is for gatherings of this kind to launch an age of real-world collaborations that will result in world-changing innovations.

This project is on a big scale, with 30 to 40 people involved, not just from Sony CSL but our OES Project partners OIST, the Okinawan electrical facility maker Okisokou Co. Ltd., SONY's Okinawan subsidiary Sony Businesses Operations Inc., and others. The scale is likely to grow from now on.

But as we have seen with other CSL research, this big-and-getting-bigger project was, 10 years ago, just a one-man research project at the lab, with no one suspecting it would swell into this great venture gathering momentum. So let me walk you through those 10 years from the perspective of a technology promoter's involvement.

3.4.3 "Packetized Electric Power": A Concept Borrowed from the Internet

The starting point was work being done by Shigeru Tajima, a CSL researcher. Shigeru wanted to create a technology that would allow everyone to easily exchange electricity with each other. One of his ideas was "packetized power"—transmission of electricity in discrete packets. Discrete packets are the form in which data travel over the Internet. Internet packets start with a header that indicates the kind of data and where that data are going, and then comes the chunk of actual data. Shigeru envisioned a system in which electricity itself, rather than data, was transmitted in the form of packets with headers attached. So he built a proof-of-concept prototype in his lab.

At TPO, this struck us as a very promising line of research but not one that a single researcher could fully exploit or one that could find a home at SONY where it could be readily turned into a product.

Shigeru's experiments demonstrated that storage of electricity was crucial to the scheme. In the case of the Internet, devices called routers briefly hold up the incoming packets and then send them on their way. To do that, they need to have the capacity to store data. For electricity packets, too, storage would be needed, meaning that batteries would be an essential component of the "Internet of electricity."

3.4.4 Making the Connection with SONY's Battery Biz

SONY was the first company in the world to bring lithium-ion (Li-ion) batteries to market. Sony Energy Device (SEND) is the business unit that makes and sells batteries. It is based in Koriyama, in Fukushima Prefecture. It seemed that the first thing we needed to do was get things moving with SEND, so I went looking for a contact person there. That's when I found out that, as luck would have it, Masashi Yasuda, who had been a colleague of mine 20 years earlier in Singapore, was now a senior manager at SEND. I got in touch with him for the first time in ages, and he consented to take a look at a demo of Shigeru's work. After seeing it, he agreed that it was a very interesting technology, and soon CSL's collaboration with SEND was underway.

Although SEND had agreed in principle to team up with us, they had their existing business to run and failed to get traction on moving the project forward (at times, 6 months would pass without getting a response from them). Sometimes at TPO, we have to give up on a project we've been trying to shepherd along because it gets completely bogged down. But when I talked it over with Mario, he suggested that I propose to SEND that we do a joint exhibit at SONY's internal technology fair.

I was skeptical; I expected SEND to simply turn us down, saying they were too busy. But to my surprise, when I proposed the joint exhibit, it sparked more active engagement on their part and we ended up doing the event. Thinking back on it now, I realize that Mario, as an engineer himself, sagaciously understood that a concrete objective such as needing something ready to exhibit at an event would spur the engineers to achieve that goal.

SEND was making and selling small Li-ion batteries to power gadgets such as mobile phones and walkmans. But they wanted to develop big batteries for home and commercial power applications, too. The "Internet of electricity" that CSL's project was aiming at would also require hefty batteries. So we embarked on a joint development effort, with SEND handling the hardware and CSL the software.

It was in late 2009, after about a year and a half of exploring many options, that the first prototype battery server (a "smart" battery) was completed. We did show it at the tech fair, but it wasn't clear where the project would go from there.

3.4.5 Putting on Public Viewings of the World Cup in Ghana

And then, flashing onto the scene like a comet, came Tsukasa Yoshimura, who I've mentioned previously. Since 2009, Tsukasa had been involved in efforts to fully leverage SONY's sponsorship of the Fédération Internationale de Football Association (FIFA) World Cup. As part of that, he embarked on a project to hold roaming public viewings of soccer matches. Although Africa has many countries with strong soccer traditions, many Africans are unable to watch their own national teams play. In Ghana, for example, only

20% of households have TV. So in 2009, the Tsukasa team outfitted a truck with a generator, projector, amp, speakers, and satellite receiver and drove it around to different villages showing soccer matches.

As soon as Tsukasa laid eyes on the battery server prototype, he wanted to take it with him to Africa.

Based on Tsukasa's in-country experience, the biggest difficulty was obtaining a reliable power supply. For example, even when holding a public viewing in a location with electricity, sudden power outages were always a possibility. (And this was especially true during the times of day when soccer matches were shown, which coincided with peak demand, at which times power would be cut off to outlying villages.) As a backup, they had a generator, but they sometimes found it difficult to obtain fuel to run it in a timely fashion. Tsukasa felt that if he had a battery server like the one developed from Shigeru's idea on site, the showing of the match would be able to continue unimpeded even during power outages. And in unelectrified villages, they could recharge the batteries from renewable energy sources, allowing them to conduct public viewings with no dependence on external sources of energy.

Meanwhile, on the R&D side, we could not have asked for a better trial run than showing World Cup matches around the backcountry of Africa. At that point, there were about six people on the development team, counting the ones from SEND, and they all jumped on the idea enthusiastically. The catch was that by then, 2009 was almost over, and the 2010 World Cup in South Africa was coming up in June 2010. Taking into account the time it would take to ship in the equipment, we had only about 3 months to get ready.

And that's not all. The team needed to build a much more robust case for the battery server than the prototype had, considering that it would have to be hauled long distances over unpaved roads in Africa. We also needed a design for the solar panel charging system and modification of the projector's power supply. Finishing all these tasks within that time frame appeared impossible. But the team brought a relentless determination and drive that enabled them to pull it off, with everything ready to go in June.

During the 2010 World Cup, the team was able to hold public viewings of the matches in 14 different unelectrified villages in northern Ghana (Figure 3.10). About 4 hours of daytime sunlight was enough to fully charge the battery server. Then it could run the equipment for two and a half hours to show the match. When the national team of Ghana scored a goal during one of these public viewings, there was celebratory bedlam, with a deafening roar of cheers. The project was a huge success.

FIGURE 3.10
(a) Public viewing of soccer in Ghana, (b) team leader Tsukasa Yoshimura, and (c) making preparations. (Courtesy of Sony Computer Science Laboratories, Inc., Tokyo, Japan.)

3.4.6 Launching a Mobile Phone Charging Service for Unelectrified Areas

The World Cup public viewings in Ghana became famous inside and outside SONY, helping research into energy to pick up speed. SEND went on to turn the battery server into a product that was marketed to government and corporate buyers and became a big business. But Sony CSL's goal was not to turn the battery server into a product, it was to create an "Internet of electricity." And that goal has yet to be achieved.

For TPO, this project gave us plenty to do in the way of public relations for external consumption and lobbying inside SONY. In order to turn OES into a real business, Yoshiichi Tokuda, who had been central to the soccer project in Africa, was reassigned to CSL from SONY corporate.

He took the lead over the next 3 years in field testing the battery server through a project with the Japan International Cooperation Agency tackling "off-grid electrification for unelectrified areas" in Ghana. They targeted the "base of the pyramid," a large but underserved segment of society.

The headline-grabbing public World Cup match viewings certainly were a turning point for serious investment in energy research at SONY. But since the equipment that traveled around Ghana actually generated and consumed its own power, the initial concept of exchanging electricity among consumer/producers was not realized. This led to the idea of a next step in field testing: recruiting village entrepreneurs to partner in offering an ongoing energy service by using solar panels to produce power and energy servers to store it and distribute it to villagers.

Yoshiichi and his team surveyed villagers and asked them what they would use electricity for if it was available. The number one answer was charging mobile phones. That may seem surprising, but an unelectrified village does not mean villagers without electronics. Many rural Ghanaians do have a mobile phone. The team asked people how they were charging their phones in the absence of electricity. The answer was: Every few days, I ride a motorcycle to the nearest electrified village, 10 kilometers away, pay about a dollar for a charging service there, and then ride back home. What a hassle! Why would it even be worth so much trouble just to have a mobile phone, you may wonder. As it turns out, a phone enables farmers to check city market prices for crops in real time, which helps them negotiate with the traders who come to buy their crops. So a mobile phone is practically the difference between life and death for a farmer, and the need for a less cumbersome way of charging it is great.

FIGURE 3.11
(a) Research activities in Ghana. Public viewing making use of renewable energy. (b) Mobile phone charging service experiment making use of renewable energy. (Courtesy of Sony Computer Science Laboratories, Inc., Tokyo, Japan.)

Our team proposed the system to local entrepreneurs and got mobile phone charging services up and running (Figure 3.11). Yoshiichi made the grueling 40-hour round-trip flight between Japan and Ghana via Dubai every couple of months for 3 years as he developed and shipped components needed in-country, conducted local negotiations, ran field tests, and filed a final report. Having delivered this innovative new energy system for developing countries out of CSL research, the same approach is now being pursued in Bangladesh.

3.4.7 Forming a Consortium for the "Internet of Electricity"

As we looked for a context in which to conduct larger-scale research on exchanging electricity between houses, we found OIST, as mentioned earlier, and with them we embarked on a joint research project in Okinawa. OIST is a recently established institution of higher learning that occupies a large campus in an idyllic place called Onnason. The campus includes not only academic buildings but also student apartments and faculty housing as well as stores, dry cleaners, and other businesses to support daily life; it is like a small town unto itself.

In a cluster of single-family homes on campus that are owned by the school and occupied by faculty and their families, each house is equipped with solar panels and energy servers and is connected to the other houses using a DC microgrid that permits the exchange of electric power. The system has been designed to enable this small community to rely mainly on renewable energy (Figure 3.12).

(a)

FIGURE 3.12
Renewable energy experiments in Okinawa. (a) Battery server.   *(Continued)*

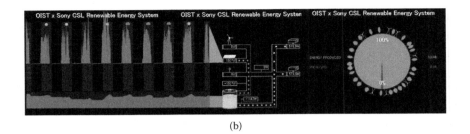

(b)

FIGURE 3.12 (CONTINUED)
Renewable energy experiments in Okinawa. (b) Visualizing renewable energy. (Courtesy of Sony Computer Science Laboratories, Inc., Tokyo, Japan.)

The project was selected to receive funding in 2013 under the Okinawa prefectural government's Subtropical and Island Energy Infrastructure Technology Research Subsidy Program. For this purpose, a consortium of Sony CSL, OIST, Okisokou Co. Ltd. and Sony Business Operations Inc. was put together and began building a platform for real-world trials of OES (Figures 3.13 and 3.14).

(a)

FIGURE 3.13
Open Energy System (OES) project in Okinawa. (a) Faculty housing hosting the platform. *(Continued)*

(b) (c)

FIGURE 3.13 (CONTINUED)
Open Energy System (OES) project in Okinawa. (b) Energy server and (c) solar panels installed in houses. (Courtesy of Sony Computer Science Laboratories, Inc., Tokyo, Japan.)

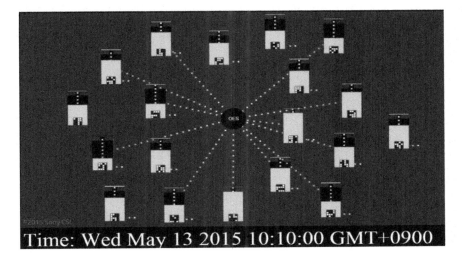

FIGURE 3.14
Power exchange system in Okinawa. (Courtesy of Sony Computer Science Laboratories, Inc., Tokyo, Japan.)

The Okinawa project required enormous efforts: putting together the grant application, negotiating with the prefectural government, coordinating among consortium members, and actually building out the technology platform—not to mention holding sessions to educate the residents of the homes. Yoshiichi was an indefatigable force in carrying through all these things, flying to Okinawa almost weekly. In order to strengthen the project's organizational structure within Sony CSL, it was formally established as the OES Project and Mario Tokoro himself, having stepped down as president of CSL, took personal charge as project leader. Yoshiichi, who had accomplished so much under the auspices of TPO, was reassigned to the OES Project.

Since then, and bearing in mind the long-term prospect of a spin-off, the OES Project has made strides to build on its early successes and progress toward changing the world!

Even before OES, TPO had been involved in supporting researchers by finding avenues for commercialization of all kinds of research. But with the OES Project, we went from serving an individual researcher at the start to becoming directly involved for the first time in executing the research itself and driving the project forward. That brought our role to an entirely new level. Although OES still has a long way to go, TPO can reflect proudly on its role in helping a tiny seed of inspired research grow into a flourishing entity that is branching out and reaching for the sky.

4
Techniques for Technology Promotion

Now that we've looked at a number of case studies, I'd like to go into more depth about what the work of a technology promoter consists of.

The work flow of technology promotion is basically the same as other kinds of sales. It starts with cataloging research discoveries, then moves on to creating sales collateral for them, followed by selling (Figure 4.1).

What a technology promoter sells is research discoveries. When research progresses to a stage of producing a new invention or new knowledge, we collate the academic papers, slide decks, videos, and other presentation materials the researcher has created into sales collateral that potential customers can easily understand. Then we go to the business units that are our customers and pitch to them. Let's look at each of these activities in more detail.

4.1 CATALOGING OF RESEARCH DISCOVERIES

4.1.1 The *Review Talk*

Researchers, having created new scientific knowledge or inventions through their observations, hypotheses, experiments, and so on, codify their discoveries, obtain patents, submit scholarly papers, and present at conferences. Then they sum all that up in a briefing before the entire Sony Computer Science Lab (CSL) staff, including management.

This briefing about research discoveries is known at Sony CSL as the *review talk*. It's held every April. Each researcher gets about 40 minutes to give a presentation on his or her research and then must submit to colleagues' penetrating questions during an open question-and-answer session. Since the achievements presented at this gathering are weighed in

FIGURE 4.1
Technology promotion workflow. (Courtesy of Sony Computer Science Laboratories, Inc., Tokyo, Japan.)

salary negotiations for the following year, every researcher strives hard to present their work in the most compelling and convincing way.

Obviously, Technology Promotion Office (TPO) staff also attend the review talk and learn a great deal. But in many research fields, the presentations turn into very abstruse discussions, so this briefing alone doesn't enable us to gather all the information we need. For example, Takashi Isozaki is doing cutting-edge research in data analysis, so his slides are filled with equations and graphs. We on the TPO team can manage to follow along with that part more or less but not the discussions, which is why right after the review talk, TPO holds follow-up meetings individually with each researcher along the following lines.

4.1.2 The TPO Interview

The most important activity in the research cataloging process is the TPO interview. Once a year, after the review talk, we spend an hour with each and every researcher. Using Takashi as an example, let's walk through the general pattern of these discussions.

TPO: Your review talk seemed pretty interesting, but parts of it were extremely technical and hard to follow. Could you explain it for us again in layman's terms?
Takashi: Sure.
TPO: At this review talk, you didn't mention [topic X] from your presentation last year. Could you update us on the status of that?
Takashi: I haven't really taken that any further, so I didn't present on it this year. Here's the analysis I'm working on right now.
TPO: Fascinating! Please let us know if that leads to anything.
Takashi: By the way, how is it going pitching to [business unit Y]?
TPO: They're doing a reorg, which has delayed their response, but they're still motivated. Next time I go there I'd like to take you along.

We spend at least an hour, sometimes more, having this kind of discussion with each researcher.

In contrast to the review talk, which mainly spotlights the theoretical aspects of research, the TPO interview zeroes in on commercialization and application prospects—identifying domains where the researchers' discoveries are most likely to find practical use and what steps are needed to move them out of the lab and into the world.

An essential part of these interviews is an Excel spreadsheet called the Technology on Promotion (TOP) List, shown in Figure 4.2.

The TOP List works like an agenda/tickler file for these interviews. We use it to guide the discussion and update it based on new things we learn from that discussion. In other words, we start with a list from the previous year and, at the end of the interview, we have a new list for the current year. It breaks down the researcher's activities by project, with fields for the project name, collaborating researchers, outline of what the research is about, year started, status of the research, and status of promotion efforts for the research.

For obvious reasons, the review talk focuses on the research areas in which a researcher has made the most progress, so it's common for research that hasn't advanced much, or is a small slice of the researcher's overall activity, to be left out. But it's always possible that one of these out-of-the-spotlight research projects could hold just the technology that some business unit at SONY needs at the moment or some application the researcher hasn't even thought of. So we are very dogged in making sure that the TOP list catalogs everything. This interview is also a valuable occasion to provide feedback to the researcher on what TPO is doing with

Technology on Promotion List
IL Alexis Andre

	Name	Developed by	Abstract	Year	Reseach Status	Promotion Status	Remarks
1	音蟲	Alexis	音が触れて見えて、再生できるシステム。	2010.2~	研究としては完成。	Open HouseとSIGGRAPHで発表。2013年 Android版作る予定 ⇒ まだ作れていない。Android側の制約	コンセプトとしては完了しているがUXを試す「場」としては継続応用予定
4	ゲーム制作	Alexis	ゲームに使われている技術の理解のために自らゲーム制作	2008.10~	継続研究中	Online Experimentsとしていくつか公開済み。継続して公開していく予定。 ⇒ ぼつぼつ製作中。	CSL HPでの公開は当面見送る方向(開発期間より公開作業のほうが時間も手間もかかるため)
7	HP Visualization	Alexis	可視化プロジェクト	2009.7~	継続企画中	CSL HPで実装済 ⇒ 25周年については終了。 新たにNew Yorkに向けて準備開始予定。	
8	LOGO	Alexis	ロゴでないロゴ	2011~	継続企画中	バイトが制作希望とのこと	8月完成の予定 ⇒ 2013オフサイトで発表 完成とは言えない状況。
9	Energy Visualization	Alexis	可視化プロジェクト@OIST	2012~	継続研究中	2012.5.18から稼働していたが、一端、撤退。 ⇒ その後、復帰。アネッタさんのプログラムのvisualizationをしている	
12	12 tone analysis	Alexis	12音解析技術と可視化技術を組み合わせたソリューション(ゲーム含)を考えるプロジェクト	2012~	終了		
14	大崎2Fプロジェクト	Alexis	Interactive aesthetics	2012~	完了	展示継続中	
15	Ars Electronica	Alexis	Interactive aesthetics	2013~	展示目標	詳細未定。だが計画はある。	

FIGURE 4.2
Technology on Promotion (TOP) List. (Courtesy of Sony Computer Science Laboratories, Inc., Tokyo, Japan.)

the various research projects on the TOP list, what business units may have a need for them, and so on.

Incidentally, I always close these interviews with the question, "What are you hiding from us?" in the style of a potboiler gumshoe. Since CSL researchers are pursuing multiple lines of inquiry simultaneously, as well as having topics in mind to tackle next, they never show their full hand at a full gathering such as the review talk. With these projects, the researchers themselves don't yet know if they will pan out or not, so they don't want to jump the gun by talking about them. But during the TPO interview with just two or three other people in a small room, we can usually get them to fess up "off the record." It's kind of like the clichéd line the informant always says to the detective: "You didn't hear it from me!"

At that point, we on the TPO side can offer our input to the researchers: "Since you're working on such-and-such, it could be interesting to combine that with this technology they have over at the so-and-so business unit…"; "I know someone with a sideline interest and expertise in that area that you might want to talk to…"; and so on. From these conversations can eventually emerge big ideas and major technologies to push, so TPO attaches great importance to these meetings.

Another response I often get when I ask, "What are you hiding from us?" is a researcher's offhanded confession, "Well, I did build such-and-such, but it's not a big deal." Researchers sometimes construct a device that they need to conduct their research or for sheer love of tinkering. And sometimes down the road that device can turn out to have relevance in some unexpected technology direction. It's another reason for TPO to give researchers "the third degree" and wring every scrap of information out of them in these interviews.

These interviews are conducted annually in April with every researcher but also on an ad hoc basis at any time of year when the need arises.

4.1.3 Multiple Information-Gathering Channels

Along with the annual review talk, twice-a-month general meetings are held at which presentations, monthly updates, reports on conferences attended, patent application status updates, and other such information is disseminated. TPO devours this and puts it to good use. In fact, I'd go so far as to say that TPO staff read what is handed out at general meetings more carefully than anyone else at CSL! This makes sense since knowledge of everything happening in the lab is our bread and butter.

But even more than these formal distributions of information, we depend on daily communication with researchers. Sony CSL promotes an environment of open communication. With the exception of when someone is in crunch time for a paper submission deadline and so on, the general rule is that everyone keeps their office door open. In the middle of the floor is the lab watering hole stocked with free coffee and lined with sofas. This is always a good place to run into people and strike up a discussion. From these daily conversations, above all else, come the seeds of technology promotion opportunities. TPO must never become some organization divorced from CSL but must stay an integral part, and these casual everyday conversations are vital to maintaining that shared heartbeat.

Incidentally, the most common words that come out of my mouth are, "Finish anything?" When I drop by the offices of various researchers on a whim, that's what I invariably blurt out unceremoniously. I'm not asking about anything specific. I just want to know whether the researcher has recently finished *whatever*: a new prototype or experiment, or algorithm, or theory. Whenever a researcher does have something fresh-baked, they start telling me about it with an excited gleam in their eyes. My "finish anything?" chats are the secret to finding out about the latest discoveries in the lab as early as possible.

For example, Shunichi Kasahara is now doing research on a head-mounted display (HMD) called JackIn (Figure 4.3). He has some new video or demo ready to go almost every time I drop in to see him. The other day when I strolled in the door and asked, "Finish anything?" he had just completed a demo on the HMD of being beaten up by a kickboxer. He had me try it out on the spot, and it was terrifying!

FIGURE 4.3
JackIn Head. (Courtesy of Sony Computer Science Laboratories, Inc., Tokyo, Japan.)

80 • Out of the Lab and on the Market

4.2 DEVELOPING SALES COLLATERAL

4.2.1 A Sales Sheet That Zeroes in on "What Does It Do?"

TPO takes the wide variety of information we gather and turns it into sales collateral. Figure 4.4 is an example of an explanatory PowerPoint slide. Our rule of thumb is one slide per technology.

That slide starts with a *technology summary* that describes in a few words what the technology is about. You can think of this in terms of the tagline for an ad. In this example, CyberCode is described as a two-dimensional barcode-enabling augmented reality.

Next we have the *technology features*, usually around three concise bullet points about what defines this technology. Each point covers a technical feature, a differentiator from similar technologies, and so on, and has a corresponding image to aid in understanding.

At the bottom of this slide is a video camera icon that links to a video that was made in the course of the research. When we do actual technology pitches, we try to show videos or functional demos whenever possible,

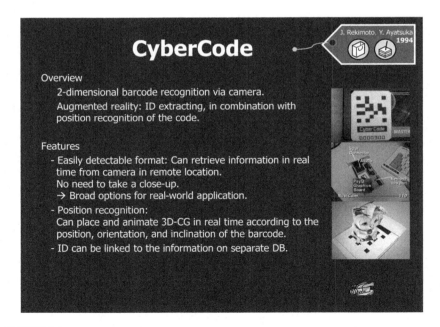

FIGURE 4.4
Technology explanation (CyberCode). (Courtesy of Sony Computer Science Laboratories, Inc., Tokyo, Japan.)

rather than just reading text off a slide, because it conveys more information in a shorter time. These materials are like sales sheets that present the essence of the technology in a nutshell.

When we have a researcher with us to explain their technology themselves, they usually start with a history of their field, describe research that went before, and then talk about what new innovation they have brought to the field. These extended presentations usually run about an hour. They are valuable for potential customers with an established interest in a technology, but for those hearing about it for the first time, they contain too much extraneous information. Another way to look at it is that having the researcher present in person is mainly for explaining "how does this technology do what it does" whereas when TPO presents, we focus just on "what does it do."

4.2.2 Technology Tag = Schematized FAQ

One characteristic of these slides is the graphic in the upper right that looks like a sales tag. We call these technology tags. They are essentially schematized FAQs.

Inside the tag is the name of the researcher and the year of the key paper or patent associated with the technology. The color of the tag indicates whether it is at the Tokyo lab, the Paris lab, or whether it has already been transferred somewhere and is now on follow-up status.

Four different icons may appear in the tag (Figure 4.5).

- The first icon is a lock, which means the research is still a company secret.
- The second icon is a paper bag, which indicates the technology has already been commercialized as a product or service.
- The third icon shows a bite of cake (which one often encounters at Japanese supermarkets offered as free samples) and indicates that some form of sample of the technology (such as trial software) is available. With software technologies, it is often the case that customers won't believe its effectiveness until they can evaluate it using their own data, so offering a sample version helps make the evaluation happen.
- The fourth icon symbolizes collaboration, which indicates that some form of collaboration is already underway—be it a case of the researcher and corporate research and development starting to work together or a case of a transferred technology now on follow-up status.

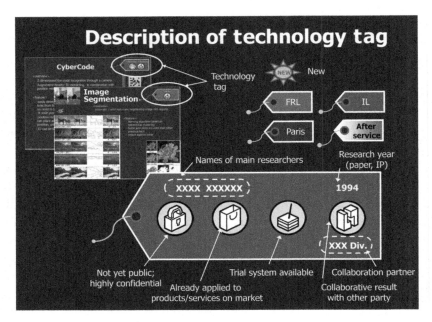

FIGURE 4.5
Technology Tag. (Courtesy of Sony Computer Science Laboratories, Inc., Tokyo, Japan.)

Along with this basic information, the slide has links to the relevant videos and websites. It's a package of all the minimum necessary information about the technology.

TPO has been in operation for a decade, and our slide deck containing a compilation of all the research we've covered over the years runs to 245 pages (as of this writing). This format has been honed in the course of pitching research discoveries to nearly 5,000 people over those 10 years. We came up with the technology tags to provide answers to the most frequently asked questions in a schematic form. Now these slides serve TPO as a sales tool, distilling discoveries down to their essentials for pitching purposes.

4.3 SELLING

The sales tools that TPO puts together as described earlier are deployed in our sales efforts in the field. Selling is the core function of TPO, the thing that makes successful technology promotion to commercialize research discoveries possible.

4.3.1 Demo Road Show: Empowering TPO to Make Initial Pitches Independently

Our main sales vehicle is the demo road show (Figure 4.6). You might have an image of the typical technology demo as a bunch of engineers showing up to see a prototype put through its paces. But our demo road shows have a slightly different style.

First of all, we are proactive. We do the typical research lab tech fairs, like our biennial open house, and take part in SONY-wide tech summits. But the results you can expect from activities like those are limited. At the open house, naturally, researchers announce their latest research discoveries. But from a commercialization standpoint, the latest discoveries are not necessarily the technologies the market needs right now. It may well be research completed 3 or 4 years ago that harbors the technology a business unit needs today.

A leading-edge research organization such as Sony CSL is aiming to make completely original discoveries, so the work done here is often ahead of the curve of the current state of the market. It is older discoveries that the state of the market has had time to catch up with that are likely to coincide with what business units are looking for, but these don't get any play at the research open house.

Moreover, design leaders and product development leaders of business units are without exception insanely busy people. Many have no time to spare to attend even an occasional research open house.

So TPO targets the key people and proactively pursues the chance to show them our demo. These key players are always thinking about the next generation product, so they are in fact interested in new technologies and will carve an hour out of their schedules. So we aim to make the very

FIGURE 4.6
Demo road show. (Courtesy of Sony Computer Science Laboratories, Inc., Tokyo, Japan.)

most of that hour, swiftly cycling through about 10 different technologies, spending 5 minutes on each.

Since it's not feasible to go into the nitty-gritty of a technology in such a short amount of time, we hone our pitch to focus on "what does it do" rather than "how does it do what it does." And we use videos and prototypes as much as possible to make the most effective use of the short time.

Another priority for these demos is making sure that TPO personnel can do them on their own. In this setting, pulling in the researchers would provide assurance that we could field unexpected questions but also attenuates the imperative for the salespeople themselves to be fully prepared. A technology promoter should not only understand each technology but have secured from the researcher enough information to answer questions that can be anticipated and be able to explain without needing the researcher right by their side.

This is probably the single-most important point in making technology promotion happen. The reality is that sometimes a demo road show scores a win, and sometimes it doesn't. Because we are doing proactive selling, sometimes right off the bat we run into an attitude of "Thanks, but no thanks; got all the technology we need right here!" Or even, as has happened to me, people will start leaving in the middle of a demo, one after another, so that by the end there is only one person left in the room. (And I think he wanted to leave too but stayed because he felt bad for us.)

However, 2 or 3 years down the road, when that business unit is drawing up its medium-term plan, we might get a request to come in and demo for them. We ask the customer to try to assemble around 10 people for the demo—not just those we are specifically targeting but anyone who has an interest.

In a way, our demos are a lot like a stand-up comedy act just getting started on the club circuit. We're the ones asking to be given a shot on stage, and we are usually presenting our demo to complete strangers—people with no preexisting interest in our technologies who, in fact, are often quite skeptical that we have anything relevant to offer them. Accordingly, success or failure rides on how we can grab the crowd's attention, win them over, warm them up, and keep them awake. We put a lot of humor into our demo pitch scripts for that very reason; we make a concerted effort to convey each technology's features with verve and panache.

Of course, we're *not* actually comedians, and we don't spend all our time writing new gags, so sometimes we let our routine get stale. I've been pitching to a certain customer a second time around only for someone in the audience to say, "Same joke as last year!" Very embarrassing! So we do refresh our patter with fresh material every year so that we can keep on getting gigs to give our demo.

4.3.2 *T-pop News*: An E-Mail Newsletter for People Who Have Attended Our Demos

When we do a demo road show, we also promote an e-mail newsletter that TPO puts out monthly called *T-pop News*. A Japanese edition and an English edition are distributed exclusively within SONY (Figure 4.7). The newsletter talks about research and technologies at CSL and reports on conferences attended, events, and coverage in the media. *T-pop News* has been published for 10 years now and has reached more than 120 issues.

There is a crucial policy for this e-mail newsletter: We only send it to people we have met face to face at a TPO demo. The reason is that nobody will ever look at some e-mail newsletter coming from a person or organization they aren't familiar with; it will go straight into the trash folder unread. So we are committed to making this a premium e-mail newsletter available only to people who are aware of what Sony CSL has to offer.

Moreover, since the content can be fairly technical, we don't expect everyone to read it. But we believe that sending it out monthly anyway maintains a sense of being in touch that may lead a recipient to turn to TPO someday when they are hunting around for a new technology.

A research lab and a business unit operate on different calendars. For the business unit, crunch time comes before the release of a new product. For CSL, researchers are most under the gun when they have a deadline to produce a paper. The e-mail newsletter is a means of loosely maintaining a tie between these organizations of very different character.

In many cases, the e-mail newsletter leads to TPO being asked in to do an additional demo road show. The newsletter and road shows work in conjunction as a system with a growing number of registered contacts.

Currently, there are about 1,500 recipients of the newsletter (Figure 4.8). In other words, because of the policy I just described, at least that many people have showed up at one of our demos over the last 10 years. (Of course, there are individuals who have been to a number of demos, and those who attended but did not wish to be signed up for the e-mail newsletter, so I estimate the total number of aggregate attendees to be about 5,000.)

Dear all,

Thank you for your continuous support extended toward our activities at Sony CSL.

For this issue, we continue New York Symposium series and asked our Deputy Director Prof.Rekimoto to write article and he wrote about hot topic, "Singularity". In the second half of this issue, Dr.Funabashi reported 2 conferences of agriculture which are held in Africa.

----------T-pop News No.117----------------

=== The Augmented Human? The Replacement of the Human? ===
In March 2015, the Augmented Human 2015 International Symposium was held in Singapore.[1] I served as one of the Program Co-Chairs and also as the organizer of the panel session titled Augmentation and/or/xor Singularity, and the Future of Human Augmentation [http://asg.sutd.edu.sg/ah2015/program]).
The symposium took an interdisciplinary approach to the augmentation of human abilities using various technologies. Discussion focused on developments in human-computer interaction, robotics, brain-machine interface, wearable computing, prosthetic technologies (prosthetic engineering), and other fields.   *snip*

[Conference Report West Africa Networking Forum 2015, ECHO East Africa Symposium (Burkina Faso, Tanzania)]
I went to Africa to see the agricultural practice, local ecosystems and attend 2 conferences in Burkina Faso and Tanzania. I hereby share an overview of the conferences.
I went to Ouagadougou in Burkina Faso to attend West Africa Networking Forum 2015 organized by ECHO, an international organization that aims to reduce hunger and improve the lives of small-scale farmers worldwide. They operate in more than 165 countries worldwide and the west africa center is located at Ouagadougou.   *snip*

[Notice]
A new company, Sony Global Education, Inc. (SGED) was established as of April 1, 2015.
The operation of application services including "Global Math Challenge" and "MathNative" will be taken over by SGED from Sony CSL. https://www.sonyged.com/en

What is the CSL T-pop (TPO Partner) Network?
T-pop is proposed as a new attempt, connecting research and business.Until now, business and research did not know where to go to discuss their mutual needs such as, from the business side, "Where should we look for this kind of technology for our next product"s function?", or from the research side "We have produced this new technology. Is there any interest?". Hence, we conceived of making a network for casual discussion.

We will release an e-mail magazine each month with the information on our offers from CSL.

FIGURE 4.7
"T-pop News No. 117" (English and Japanese) 2015. (Courtesy of Sony Computer Science Laboratories, Inc., Tokyo, Japan.)   *(Continued)*

＜T-pop 記事の例＞

T-pop の皆様へ

ソニーコンピュータサイエンス研究所　テクノロジープロモーションオフィスの夏目です。
今回は、NY シンポジウムシリーズで、暦本副所長から今話題の Singularity について寄稿頂きました。後半は、舩橋研究員が参加したアフリカでの農業に関する二つの学会報告です。

------ T-pop News No.117 ------

【THINK EXTREME】
=== 人間の拡張？人間の置換？ ===

この三月に、シンガポールで Augmented Human 2015 という国際学会が開催された[1]。私は学会運営者としてプログラム委員長を務めると同時に、"Augmentation and/or/xor Singularity, and the Future of Human Augmentation" と題するパネルセッションをオーガナイズした。

Augmented Human は、人間の能力を拡張する諸技術に関する成果を議論する学会　で、Human-Computer Interaction, Robotics, Brain-Machine Interface, Wearable Computing, Prosthetic Technologies (補綴工学)などの成果が発表される学際的な場である。……（後略）……

【学会報告（ブルキナファソ、タンザニア）】

アフリカにおける農業慣行や生態系の観察、国際会議に出席するためにブルキナファソとタンザニアに行って参りました。以下、会議の概要についてご報告させて頂きます。

*West Africa Networking Forum 2015

West Africa Networking Forum 2015 (西アフリカネットワーク・フォーラム 2015)に出席するためにブルキナファソの首都、ワガドゥーグーに行って参りました。……（後略）……

[お知らせ]　書籍出版

CSL 社長　北野さん企画・執筆による書籍が羊土社より出版されました。
『Dr.北野のゼロから始めるシステムバイオロジー』……（後略）……

CSL T-pop(TPO Partner) Network とは：

T-pop は、研究とビジネスを繋ぐ新たな試みとして提案しました。ビジネスサイドから　「次の商品やサービスの機能としてこういう技術を探している」とか研究サイドから「こういう新しい技術ができたので、興味があるか。」など、今までお互いにニーズがありながら、どこに聞いたらいいか分からないというようなことを気軽に話せるネットワークとしていきたいと考えております。CSL 側からは、月一のペースで、メールマガジンを流させて頂きます。……（後略）……

FIGURE 4.7 (CONTINUED)
"T-pop News No. 117" (English and Japanese) 2015. (Courtesy of Sony Computer Science Laboratories, Inc., Tokyo, Japan.)

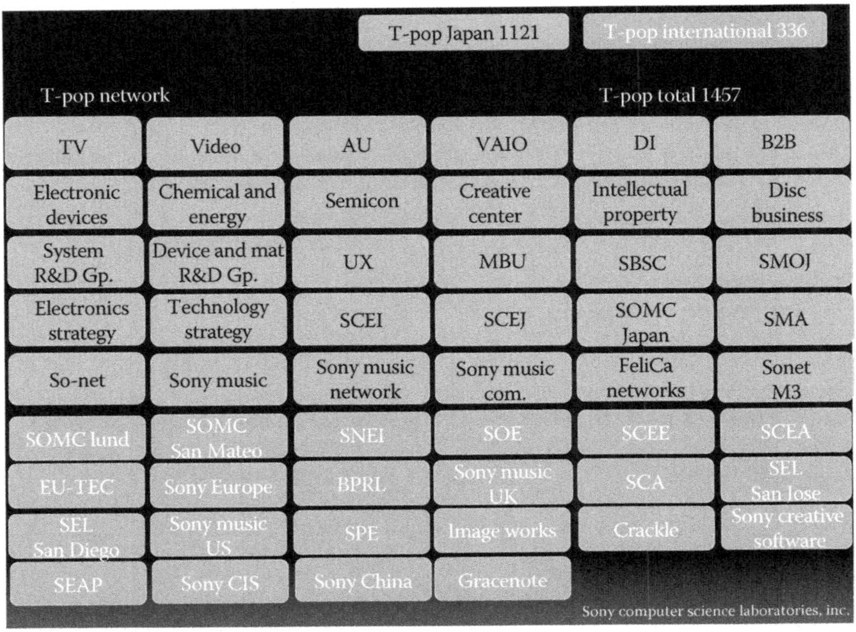

T-pop network			T-pop Japan 1121		T-pop international 336	
					T-pop total 1457	
TV	Video	AU	VAIO	DI	B2B
Electronic devices	Chemical and energy	Semicon	Creative center	Intellectual property	Disc business
System R&D Gp.	Device and mat R&D Gp.	UX	MBU	SBSC	SMOJ
Electronics strategy	Technology strategy	SCEI	SCEJ	SOMC Japan	SMA
So-net	Sony music	Sony music network	Sony music com.	FeliCa networks	Sonet M3
SOMC lund	SOMC San Mateo	SNEI	SOE	SCEE	SCEA
EU-TEC	Sony Europe	BPRL	Sony music UK	SCA	SEL San Jose
SEL San Diego	Sony music US	SPE	Image works	Crackle	Sony creative software
SEAP	Sony CIS	Sony China	Gracenote		

Sony computer science laboratories, inc.

FIGURE 4.8
Where mail magazine is distributed. (Courtesy of Sony Computer Science Laboratories, Inc., Tokyo, Japan.)

The list includes a far-flung network of people from Sony Music Entertainment, Sony Pictures Entertainment, Sony Life Insurance Company, PlayStation game developers, and more. Before the establishment of TPO, very few people in those corners of SONY would have even heard of Sony CSL. So I take pride in the extensive network of interested colleagues we have built.

4.4 THE IMPORTANCE OF THE MoTR

When TPO uses these various means to generate interest at a SONY business unit, put them in touch with the researchers, and transfer a technology, we draw up a Memorandum on Technology Release (MoTR) (Figure 4.9). This is a document that CSL and the business unit sign when the necessary technology information, software, and so on are actually handed over. I touched on this earlier in my recounting of technology transfer case studies. When transferring a technology, the business unit might express appreciation and promise to give CSL credit when it

Memorandum on Technology Release

Sony Computer Science Laboratories (CSL) is ready to release its technologies to other groups within Sony in order to contribute to Sony business. We request each Receiver to agree on the following conditions beforehand. As this is not a legal document, we do not request for signs. Instead, we create Time-Stamped PDF files to keep at respective sides, after achieving mutual consent on the conditions described hereunder.

1. **Purpose of Technology Release**
   - The purpose of the technology release from CSL is to contribute to Sony Group business, directly or indirectly. The Receiver may utilize technologies for both short term and long term objectives inclusive of further study and development.

2. **Confidentiality**
   - Both parties shall maintain the confidentiality of the information disclosed as confidential, whether orally or in writing. Both parties shall not disclose such information to any other parties outside the respective divisions in charge.

3. **Limitations on Use of Released Technology and its Deliverables**
   - Without prior consent of CSL, the Receiver shall not use CSL technologies, the deliverables and/or derivative results originating from the released technologies, for purposes other than those defined at the timing of technology release.
   - Without prior consent of CSL, the Receiver shall not re-distribute CSL technologies, the outcome and/or derivative results originating from the released technologies to any third parties or divisions, outside their own.

4. **Cost Sharing**
   - Basically CSL does not request for any license fee on the released technologies.
   - However, with prior consent of the Receiver, CSL may request the Receiver to cover the actual expense related to the technology release, and/or additional development work requested to CSL related to the requested release.

5. **Limitation of Liability**
   - The Receiver shall not deem CSL liable for any damages or losses resulting from applying the released technologies or information to the Receiver's projects.

6. **Request for the Credit-Line**
   - CSL requests that the origins of the released technologies are appropriately credited in case of demonstrations / announcements (both internal and public) of the deliverables and/or

FIGURE 4.9
Technology transfer memorandum. (Courtesy of Sony Computer Science Laboratories, Inc., Tokyo, Japan.) *(Continued)*

derivative results of the released technologies.
- In case of internal demonstrations / announcements, please indicate the "name of researcher, laboratory name (e.g. Tokyo, Paris), Sony Computer Science Laboratories Inc., ". Furthermore, please indicate the type of collaboration - whether "joint research", "principle research", "technology supply", "concept supply", "IP contribution" etc. CSL may ask the Receiver to include "CSL Collaboration Label" in panels or brochures. Following are example of internal demonstrations / announcements:
  - STEF
  - Open House of each group
- In case of public demonstrations / announcements, please consult CSL in advance for details. Early discussion is advised, especially if Trademarks are involved.
  - Press Release
  - Display at new product announcement
  - Exhibitions at shows
- CSL retains the right to appeal the accomplishments of the technology release after the Receiver has achieved the target of the technology transfer (both internally and publicly). In this case, CSL will state that the result is a collaborative effort between the Receiver and CSL.

7. **Request for Feedback**
- CSL may request for feedback from the Receiver inclusive of non-technical matters, suggestions, level of market success, voices of customers, etc. so as for the betterment of CSL's activities.

8. **Technologies to release**
   -

9. **Persons in Charge of this Memorandum**
   XXXX

   E-mail:
   CSL   Tetsu Natsume
               Technology Promotion Office, Sony Computer Science Laboratories, Inc.
   E-mail: CC:

FIGURE 4.9 (CONTINUED)
Technology transfer memorandum. (Courtesy of Sony Computer Science Laboratories, Inc., Tokyo, Japan.)

appears in a product someday, but sometimes a long time passes before that occurs, the personnel at the business unit get shuffled around, and the promise is forgotten. In fact, sometimes the very fact that a technology came from Sony CSL gets dropped from its documentation. That is why we arrange to get the promise officially, on paper and signed, and why we continue to track the technology after it is transferred out of CSL. The MoTR is the instrument we use.

This document covers handling of confidential information, allocation of costs, and other matters that must be dealt with in executing a technology transfer. We have fine-tuned the memorandum so it can be wrapped up quickly. First of all, we avoid analog paperwork with signatures and do everything via e-mail. Once the content of the MoTR is agreed, we create a PDF that we and the business unit will both keep copies of and that constitutes the agreement.

Furthermore, we don't insist that it be (digitally) signed off on by someone with a specific title; it doesn't have to be a department head, it can be a project leader, or whoever is directly responsible. In fact, we make every effort to ensure that the MoTR is agreed upon with the person who has immediate custody of the technology, rather than some higher-up who is likely to be rotated out of their position down the road. In contrast, the person who is actually in charge of the technology often stays in place for a long time. So we prefer getting the commitment from whoever will actually be at the controls in moving the technology forward.

4.5 PHILOSOPHY ON LICENSE FEES FOR TECHNOLOGY

Another thing we are frequently asked about is costs: "If a Sony CSL technology is used in a product or service developed within SONY, do licensing fees have to be paid?" Our response is, "It's already been paid for; there are no additional charges."

What we mean is that Sony CSL is funded out of SONY corporate's budget, and that budget, obviously, comes from the sales and profits of the business units. In other words, each business unit is indirectly paying Sony CSL already, so the research discoveries made here do not carry any additional charges (although in some cases we ask for compensation of specific expenses such as travel and equipment).

I don't think it's appropriate to charge any additional fee. It's the way of the world that people who pay money expect to have a say in things. If we were to accept money in a form equivalent to a technology license, the business unit would naturally expect a level of service commensurate with that.

And if they started trying to dictate priorities to researchers because of financial leverage, it would turn Sony CSL's values upside down.

TPO's job is to elevate the achievements of Sony CSL as a whole and make it clear to any observer that we are contributing to the company. If we can't do that, we will not be able to keep our budget. The financial contribution of CSL discoveries is an important yardstick. But we wish to rigorously avoid receiving our funding in the form of a license fee. Preserving an environment of free inquiry in research at all costs means refusing direct payments from business units to which technology is transferred.

On the other hand, in cases of collaborative research with outside companies, we can't just give away discoveries for free. There is no rigid policy about how Sony CSL should be compensated; it is handled on a case-by-case basis. But the most basic issue is whether the researcher desires to do the outside collaboration. TPO never insists that a researcher work on something they don't want to. Based on the nature of the collaboration, the cost-sharing split, rights to any discoveries, and so on are drawn up in a contract with the help of the legal and intellectual property people at SONY corporate.

In addition, we have used the method of externalizing licensing of already completed research through Koozyt. For example, with CyberCode, it was complete as a research project, and there were many parties outside SONY eager to use it. It's difficult for a research organization like Sony CSL to field a flood of licensing requests, so we licensed CyberCode to our spinoff company Koozyt, and Koozyt then handled the licensing and technical support.

We anticipate more collaborative research with other companies in the future, and TPO's capabilities will shine through in striking the right balance between creating an environment of independence for researchers and dealing with contractual and rights issues.

4.6 THE 10 CORE PRINCIPLES OF TECHNOLOGY PROMOTION

So far, we have covered a number of methodologies for technology promotion. I also explained how TPO obtains information from researchers, turns that into easily digestible sales collateral, and pitches it to various business units. And I mentioned the importance of *T-pop News* in building a network of contacts across the SONY organization.

So, is it enough to faithfully utilize these methodologies and expect successful commercialization to follow? Not at all! If that was the extent of TPO's sales efforts, we would be a total failure.

These methodologies are, to use a sports metaphor, like basic fitness training. Fitness training is essential for victory on the playing field. But if you think just being fit is enough to win, you are sorely mistaken.

Like physical fitness, appropriate sales methodologies are only the starting point. From that foundation, you must build with all kinds of daily disciplines—follow-through, adaptation, and constant experimentation—before you get results. There is no cookie-cutter strategy, and every case is different. Sometimes I use a connection to an old colleague, sometimes I cold call an organization on which I've been keeping tabs. Sometimes I take the initiative, sometimes someone comes to me. Sometimes we keep things all in the SONY family, and sometimes we go outside to find the right resources. Technology promotion requires this kind of perpetual adaptation and flexibility. From the case studies presented in previous chapters, that should be obvious.

Figure 4.10 shows four mottos that we came up with in 2006, when TPO was getting started. At that time, we believed that speed, cost, sincerity, and obsession were the four major factors in technology promotion.

Over the ensuing years, I expanded that into the following 10 core principles of technology promotion. These distill the accumulated lessons of 10 years of selling research discoveries.

FIGURE 4.10
TPO motto in 2004. (Courtesy of Sony Computer Science Laboratories, Inc., Tokyo, Japan.)

4.6.1 Principle 1: There's a Right Time to Bring Every Discovery out of the Lab

On average, it takes 10 years for a frontier research discovery to achieve commercialization. In particular, truly trailblazing research often runs far ahead of changes in the marketplace. Technology promotion requires monitoring the changing marketplace carefully and recognizing when the time is ripe for a discovery to be recalled, then making a forceful push to get it out the door. In order to make those judgments on the right timing, it's important to always have your antennae out and attuned to the direction technology is moving and changing. Meanwhile, researchers shouldn't stagnate with a particular discovery but should constantly maintain their pursuit of new discoveries (examples: CyberCode, FEEL).

4.6.2 Principle 2: Use Every Possible Connection

The success of technology promotion depends on how wide you can cast your net of business opportunities. So building a network that's not restricted to SONY but encompasses many industries and companies is essential. And you must use not only your own personal network but actively tap the networks of founders, chief executive officers, researchers, and friends of friends, as the need arises (examples: semiconductor manufacturing, Moe-Kaden, 12Pixels, Open Energy Systems).

4.6.3 Principle 3: Don't Hold Preconceptions About Other Organizations

Sometimes, before going to meet people from an organization, you have a preconception, such as "They're conservative, they won't be interested in new technology" or "Their business is based on a totally different area of technology, they won't be interested in this." But then, after actually meeting them, you find that you were wrong and there is a better fit than you anticipated. So avoid preconceptions and be open to linking up with any and all organizations that you can (examples: CyberCode, semiconductor manufacturing).

4.6.4 Principle 4: Assume That Customers Need a Simple Message Drilled in Hard

In the grind of their day-to-day responsibilities, people on the business side place a low priority on talking to researchers. Technology promoters have to generate interest and enthusiasm from the customer within a limited contact window. That is why TPO prioritizes "what can this do" over "how does this do what it does" in our pitches about technologies and sets out to present the ones that have the best potential fit.

4.6.5 Principle 5: Mental Fortitude

In sales, you will always face instances of the door being slammed in your face or of being told off as an idiot when you can't answer a customer's question. Some of the harsh receptions I've received include "well short of expectations," "what a waste of my time," and "you haven't given us enough even to consider it." If you can't endure that kind of criticism, you're not cut out to be a technology promoter.

4.6.6 Principle 6: Try Things That Haven't Been Tried Before

Commercializing research discoveries is about pioneering new approaches. In technology promotion, you are going to encounter all kinds of disagreements and resistance. To get something commercialized, you have to search out new approaches that you believe can happen (examples: Extractor Discovery System music analysis technology, Aha! Experience, Koozyt).

4.6.7 Principle 7: Move Fast! Every Second Counts

When someone on the business side who is slammed with work shows interest in a new technology, don't let that moment of opportunity slip away! But interest is a rapidly depreciating commodity. If the interest level at the moment you receive the e-mail from your contact is a 100, by the next day it will drop to 50, 2 days later it will be 30, and a week later it will be all the way down to 10. (I call this the "half-life of technology interest.") Accordingly, technology promoters need to jump on any expression of interest without a minute's delay.

4.6.8 Principle 8: Follow-Up After Technology Transfer Is a Must

After transferring a research discovery to a business unit for development, it will have to pass through numerous hands and processes before it is turned into a product, so it's never the case that we at TPO will keep on dealing with the same people we transferred it to originally. The technology promoter must track the technology wherever it goes and keep tabs on its status until it reaches the market (examples: VAIO Pocket, FEEL).

4.6.9 Principle 9: Don't Just Be a Technology Promoter, Be a Research Producer

Never lose track of the endgame: commercializing research discoveries. Always be asking yourself, "How can I get this to market?" Pitch it across a wide range of business sectors, consider trying to make it an international standard, hold events, do a deal with a photo-booth company, or hire celebrities—whatever it takes to get over the finish line. Never be passive. A movie producer does whatever it takes to get that film up on the silver screen. Be a research producer who does the same (examples: Aha! Experience, Moe-Kaden, 12Pixels, FEEL).

4.6.10 Principle 10: Never Forget That You're Doing It for the Lab

And the most important principle of all is to always remember that technology promoters are working on behalf of the lab's researchers. Don't take a detached, critical stance like a corporate-style research manager or some kind of picky arbiter of research value. You are a part of the research organization. You are a partner with the researchers in turning their discoveries into business contributions. You must never forget the responsibility you bear.

4.7 WHY TECHNOLOGY PROMOTION IS NECESSARY

I have been talking about TPO activities and a number of telling cases. I explained that our basic work flow goes from cataloging research discoveries, to preparing sales collateral, to selling. As seen through the semiconductor,

Aha! Experience, Koozyt, and Open Energy Systems cases, the term "commercialization" actually encompasses a vast diversity of methodologies.

Ultimately, there is no by-the-book approach that works for commercializing all research; it depends on the nature of the specific research in question and requires devising new and creative approaches in each instance. This kind of support is not something that can be delivered from an external source; the technology promoter has to be embedded in the research lab—part of its atmosphere. So I emphasize yet again that technology promotion must be a function integral to the research organization itself. I want to close this chapter by revisiting the reasons for that.

4.7.1 Overcoming Divergences of Timing and Setting

The role of technology promotion is to overcome divergences of timing and setting to connect research and business. As I have stated over and over, when the research is done and when the market is ready for it are two different times. In my experience at Sony CSL, I have seen plenty of research that does not get commercialized until 10 years later. In the meantime, the researcher can't spend 10 years parked on the same research topic; he or she has to keep pushing forward. On the contrary, if the research is simply forgotten, it will never be commercialized. Technology promotion's role is to bridge those intervening years.

Then there is divergence of setting to deal with. By setting, I don't mean physical location so much as organizational location. A typical corporate research lab tends to have more heavily weighted links with certain parts of the company. For example, even if a researcher is in frequent contact with the product development personnel of certain business units, he or she is unlikely to have connections with sales departments, manufacturing operations, or practically unrelated businesses such as media or finance. Overseas business units are probably even more off his or her radar. But since commercialization opportunities for research could come from anywhere, the more far-flung a contact list you create across different organizations within the corporate entity, the better your chances of succeeding in commercializing the research.

Since it's unrealistic to expect researchers themselves to build up an enterprise-wide network of connections, there needs to be a technology promotion operation dedicated specifically to building and maintaining such a network.

4.7.2 Role Division between Researchers and Technology Promoters

As the numerous examples presented here show, we aim for Sony CSL commercialization efforts to be undertaken with a division of roles between the researchers and the technology promoters. This role division can be seen not only in the field of research but in many other fields: the author/editor relationship, the film director/producer relationship, and so on. There is a tag team between the creative geniuses and those who can—through taking an objective view of the work—bring that work to the world in the most effective way.

The producer/editor role must not only understand the work of the author/director role but have the specialized expertise necessary to present it to the public or audience through choice of title, promotional efforts, and so on. The technology promoter, likewise, must be a consummate pro who understands the work of the researcher and determines how to communicate its qualities to businesspeople to find avenues for commercialization.

For any given researcher, opportunities to commercialize a technology he or she created come along only once in a while, whereas the technology promoter is involved in far more opportunities spanning many researchers and consequently has accumulated methodologies and know-how that can be effectively tapped to achieve commercialization as new discoveries come along.

4.7.3 Technology Promotion Must Be Part of the Lab

Another point that deserves to be reiterated here is that technology promoters need to be part of the lab. Research planning and research strategy functions are often performed by organizations external to the research lab itself. That creates a tendency for them to be perceived as either "supervisors" or "assistants." It makes them outsiders rather than insiders and creates distance between the technology promoters and the researchers. Those conditions do not allow for smooth efforts in commercializing research discoveries.

So it is critically important for technology promoters to be truly in the same boat with the researchers and reporting to the head of the research lab.

4.7.4 Beyond Technology Promotion

Next, there is the question of the status of technology promotion. The technology promoter is committed to seeing research commercialized.

Many research labs have a research planning unit, but research planning is, in a sense, looking down on the research itself, and it is not clear what a research planning unit is actually accountable for accomplishing. Technology promotion, as a sales organization, is responsible for making sales. It can be evaluated on the basis of whether or not research discoveries are being successfully commercialized. Of course, there is research that is easy to sell and research that is hard to sell, but the tougher the sale, the more a salesperson proves his or her mettle.

And only technology promotion requires the ability to explain research as a core selling skill. (In product sales, if you have to take a product designer along with you to explain to the customer what the product does, you're a failure as a salesperson. The same goes for technology promoters.)

But technology promotion takes more than just putting together sales collateral and pitching. Depending on the particular research, you have to pitch in different places, spin different stories of potential commercialization, and make things happen. That role is richer than the term "technology promoter" can fully encompass, and it is really a function I would describe as "research commercialization producer."

In fact, when I was considering writing this book and contemplated my actual role at TPO, some suggested that I label the job as "research producer" rather than "technology promoter."

But I am very attached to the term "technology promotion" and was determined to make it a keyword in this book. More accurately, I am attached to the corresponding term in Japanese, which actually literally translates "research sales." I believe that a technology promoter—a salesperson for research—should aspire to the category of top salespeople in more conventional sales organizations—the ones who hit the biggest numbers by doing more than just going around delivering canned pitches but rather by crafting compelling stories that sell each product. At TPO, we are proud to be salespeople and want to be among the best at it.

4.7.5 How Selling Research Changes Research

The biggest contribution that technology promotion makes to a research lab is that it changes the research itself.

The existence of technology promoters within the research lab organization itself provides an outlet for commercialization. That fosters in the researchers an awareness that the research they do won't just result in academic papers but will make its mark on the real world. They will be more inclined to write software that other people, not just themselves, will find convenient to use; they will compile information on their own research more assiduously; they will imagine how their discoveries will be used in the world; and all this will propel them on in the research they do. As a result, they will do better quality, more grounded research. Not research for the sake of doing research but research for the sake of bettering humanity. And that is the most important thing of all.

For all the aforementioned reasons, I believe that every research lab should incorporate a technology promotion function.

Section II

Researchers on Technology Promotion

How do our researchers themselves see technology promotion? Tetsu Natsume of the Technology Promotion Office (TPO) interviewed several Sony CSL researchers.

II.1 IMPLEMENTING OUTRAGEOUS IDEAS

Alexis Andre, CSL Researcher

First up is Alexis Andre, already introduced on page 17.

Alexis, I'm always grateful for your enthusiasm for TPO! Would you start off by telling us a little about your previous research work, which didn't feature technology promotion, and your research with Sony CSL?

Prior to joining Sony CSL, I was a PhD student, but after joining this organization, my attitude to research changed considerably. At the university, the final aim of your research was to write a paper. I never gave any thought

Alexis Andre, CSL researcher.

to whether or not my research could be of any use in the wider world. In fact, I had pretty much given up on such issues from the very beginning.

Why was that?
Basically, there was no way to put your ideas into practice. Even if you thought your research could be of some use, without the route to achieve that, you had no idea what to do. So even as I wrote papers proclaiming how my proposals were superior to preexisting technologies, without a way to get those ideas out there into the world, I always felt it was all in vain—even as I presented my findings.

And did you find the setup at Sony CSL to be rather different?
Very different. First of all, what Sony CSL asks of us is to realize technologies that will change the world. So even if you go several years without publishing a paper, they don't mind as long as your research is heading toward something concrete. Then, the TPO will help link you up with somewhere that might be able to put your ideas into practice, so you really feel the potential to make a positive change in the world. That's a big difference.

Of course, whether or not you will be able to implement your research depends on the actual strength of the research itself. But I think that as long as the possibility is there, it makes a big difference to the motivation of researchers.

From the perspective of your research, what has been the importance of technology promotion?
I think there are two main points. The first is that you are able to go about your research with that goal of implementation in sight. That awareness of whether or not your project can be realized has a big influence on your research. For example, let's say you're demonstrating some new software. If it's just for an academic society, it's enough for the software to work on your own personal computer. But if implementation is the goal, you have to think about passing that software on to somebody else, for *them* to use. Even if it's not for a commercial release, when you're aiming for a package that people are going to use, that influences your approach.

The second point is something I never expected: the outrageous requests we have to field. From "make a new form of energy like nothing anybody's ever seen before," which is what we did with the Open Energy Systems project; to "build us a net system that really livens up meetings" for use in TPO planning meetings; or "create a logo that doesn't look like a logo"; or "we want a manual that doesn't include any text whatsoever." These are just some of the crazy ideas people come to us with, but each one is fascinating from a research perspective. That's another way in which I'm grateful to the TPO team!

II.2 IMPLEMENTATION OF ACADEMIC VERSUS CORPORATE RESEARCH

Jun Rekimoto, CSL Deputy Director

Next, let's hear the thoughts of CSL Deputy Director Jun Rekimoto, who is also a professor at the University of Tokyo.

Jun, you've been at Sony CSL since before the introduction of technology promotion. What differences have you noticed since the establishment of the TPO team?
One difference is the way in which CSL has become more widely recognized within SONY as a whole. In the past, a lot of colleagues used

Jun Rekimoto, CSL deputy director.

to see our laboratory as some sort of ivory tower, conducting high-level research that didn't necessarily have any bearing on the business of the wider company. In other words, they didn't see the practical use in what we were doing. But since the establishment of the TPO team, you don't hear that sort of opinion anymore. And, thanks in part to TPO demos and *T-pop News*, it's clear that there are more and more people who have heard of Sony CSL, even among those with whom we have had no prior dealings.

Another thing is that it has become easier to hand over our research. Previously, we would pass successful research on to colleagues, only for it to appear in exhibitions or be incorporated into products without proper credit to Sony CSL. It used to bother us that the people we passed the baton to might take all the credit, and that our own research and effort here at CSL would be forgotten, so sometimes we would hold back some aspect of our findings or try to make sure to pay regular visits to the other groups using our research just to try and maintain our involvement. Nowadays, TPO systematically arranges the proper documents to authorize the transfer of technology, and things like the inclusion of a collaboration mark, so we don't need to go out of our way to stay involved.

As part of our technology promotion, there are ongoing collaborations with outside companies including Daiwa House. Does that have an impact on your research?
From a research standpoint, it is exceptionally meaningful to forge connections with various different industries. For example, working with Daiwa House gives us various insights on the home-building industry that in turn give birth to new ideas. It's worthwhile to broaden your horizons.

Parallel to your work here at Sony CSL, you also work as a university professor. What do you think is the primary difference between working at a university or a business when it comes to putting your research into practice?
At a university, whether or not you can actually implement your research depends to a great extent on the attitude of the supervising professor. At present, I feel like if you want to actually build on university research, you have to either enter into partnership with an outside business or spin out to form your own company through which to turn your ideas into products. But many professors have no interest in such activities. At a university, the main goals are to teach students and have them conduct research, write papers, and finally graduate. That in itself is quite challenging, so it's hard to find the mental energy to devote to other things. So, when it comes to actualizing your research, it's easier to do that as part of a company.

What is it actually like to be part of a university-business research partnership?
It's actually very tough. Apportioning rights, deciding on contracts and costs, and things like that—there are so many things you have to do before even getting started on research, and of course there's no guarantee that your work will yield the expected results. If they're not fully satisfied, the business side might refuse to take things further. But research results, rather than fitting within strict, pre-determined guidelines, tend to emerge as you gradually connect the dots within a more expansive framework. That's why technology promotion has an extensive network for linking up the *needs* of various businesses with the *seeds* generated by our research. I think that's the ideal approach—to begin a partnership when those needs and seeds match up.

To achieve that, you need an up-to-date register of all your research projects. Is that something that can be achieved at a university?
That's another problem. At the University of Tokyo, we have over a thousand individual laboratories, and each one will have a number of ongoing

projects at any given time. It would take a massive system to cover all of that. The efficiency of the TPO team at Sony CSL is thanks in part to the current small scale of our laboratories. If you wanted to do something similar at a university, it wouldn't work unless each department or faculty had its own technology promotion team.

II.3 PAPERS ARE FINE, BUT NOTHING BEATS THE JOY OF RESEARCH THAT BECOMES PRODUCTS THAT CHANGE THE WORLD!

Takashi Isozaki, CSL Researcher

Let's finish with some comments from Takashi Isozaki, who joined Sony CSL from the laboratory of another company.

You were working in research at a different company. What have you noticed?
My previous workplace was part of a corporate research division, but it lacked a dedicated system for technology promotion. While I was there, I actually spoke to division management about the need for something like that, so I was really surprised when I came to Sony CSL and found there was just such a system in place.

How did technology transfer happen at your former workplace?
There were various ways. One thing I remember is how, following one case where the lab proposed its research directly to the relevant business department, we saw an increase in demand to conduct research in accordance with business needs. Also, while research linked closely to current operations had a high likelihood of being used, it seemed harder for research into newer fields to actually get put into practice. That was exactly the kind of field I was involved in, so it was tough.

What sort of ties were there to the business division?
There were cases where someone from the research division who had moved to the business side would contact me through an old acquaintance in research about working together or developing something specific. But those methods relied on chance personal connections within the company and were not particularly systematic. That's why I felt the need at the time for a dedicated technology promotion setup.

Takashi Isosaki, CSL researcher.

At the moment, there are numerous projects in progress across the SONY Group that are based on the application of data analysis, and we at the TPO team are also contributing to that. But could you tell me what sort of influence technology promotion has had on your own research?
Data are crucial to my own research, so having access to a variety of data is incredibly useful. University professors are always complaining that they need more data, so I'm really grateful to have access to so much data from different businesses within the SONY Group. And in the process of analyzing so much data of various types, you notice recurring themes that present clear priorities for future research, which is very important.

And what have you felt about your interactions with the business division?
Recently, when presenting a new piece of technology at one business group, there was a comment that seemed to make a particular impact on my counterparts. I said: "We still write papers (i.e., we are working with cutting-edge technology), but that on its own is not enough. At Sony CSL, we see our research through until it can be of use to the world." They really nodded their heads when they heard that.

They had probably thought that, as a basic research lab, Sony CSL wasn't too interested in adapting research for the business side. I explained how that wasn't the case, and I think they changed their minds. People who work in business areas far removed from research tend to carry the assumption that they won't be able to work with researchers, so it can be hard to make a connection. That's why a technology promotion team like ours is so important, to actively make connections at the leading edge of both research and business.

Section III

The History of Sony's Technology Promotion Office (Mario Tokoro)

5
Before TPO

5.1 LAB-DRIVEN PRODUCT DEVELOPMENT: THE PRECURSOR TO TPO

In July 1987, I was an associate professor at Keio University. One day, I received a visit from Toshitada Doi, at the time head of SONY's NeWS Workstation business. He told me that he wanted me to join the company, to improve its technology in the area of computing, where SONY found itself lagging somewhat behind. I made a proposal of my own: that together we might form the world's top research facility. In 1988, following approval from SONY for this idea, together Doi and I founded Sony Computer Science Laboratories (Sony CSL), which I was to run alongside my responsibilities at the university.*

My strategy at first was to remain relatively inconspicuous within SONY, while building a solid reputation as a research lab in the outside world. The fundamental aim of a company is to generate profits, so against a rapidly changing business backdrop, I felt it would be difficult to win praise within our parent company for fundamental or leading-edge research. My aim was to contribute to the company's profits over the long term. Accordingly, in order to conduct basic or frontline research with a view to moving beyond the firm's established field of activities into new business areas, I believed it was necessary to earn outside recognition, namely among communities with the capacity to assess fundamental and leading-edge research, and to use that as a base from which to demonstrate the worth of our lab within SONY itself. That was something I had learned from my own experience of conducting collaborative and funded research with numerous companies, along with my experience inside and outside Japan.

---

* For the full story in Japanese, refer to *Tensai isai ga Sony no fushigi na kenkyujo*, Mario Tokoro and Shinko Yuri, Nikkei Business Publications, 2009.

At that time, that strategy began to earn international recognition for Sony CSL for its research into fields such as operating systems (OSs), the Internet, and user interfaces. One such project was the Apertos distributed object-oriented OS developed by Yasuhiko Yokote and others. "Distributed" is the ability to operate several computers connected by networks to work as a single computer, while "object oriented" refers to technology in which an autonomously operating program known as an object organizes software. At the time, several labs around the world were aiming to construct an object-oriented application. But we wanted to combine that with a distributed system to produce an OS. It was a highly ambitious and unprecedented project.

The intent was to incorporate Apertos into products to eventually become SONY's standard OS. With the full backing of the then president of Sony CSL Dr. Doi, and Akikazu Takeuchi as the leader, Yokote and several other researchers and engineers were transferred to SONY in 1996. Under the new name of Aperios, the software subsequently made its successful debut on the market as the OS of products including SONY's first satellite receiver (the Set-Top Box) and the dog-shaped robot pet, Aibo.

Several years later, the rise of open-source software precluded the further development of Aperios. But, thanks to the work of a team of researchers and engineers under Yokote's leadership, the technology went on to play a key role in the global domination of Sony Entertainment's PlayStation 3.

Based on work including Fumio Teraoka's Mobile Internet protocol and Atsushi Shionozaki's real-time protocol, the AMInet Project was established with a view to the efficient transmission of real-time voice and video data, and, from 1999 onward, researchers and engineers departed one by one for SONY itself as the firm sought to commercialize the technology. In this case, however, Teraoka and Shionozaki remained at CSL. Although the AMInet Project failed to generate any tangible products, it did lend momentum to the engagement of SONY's consumer electronics goods with the Internet, an area in which the firm had been lagging.

Taking a step back to survey such output, the aim toward a linear model driven by the laboratory or by research itself becomes apparent. Under such a linear model, for each revolutionary new technology developed by the laboratory, the related technologies needed for commercialization are developed in-house, and profit is generated by the sale of the various products that arise.

Back in the 1990s, hopes were high for steady economic growth, and businesses' growth strategies seemed to be the right approach. In line with such thinking, the strategies behind the Aperios and AMInet can be seen as appropriate for the times. With such technology and product development as its core competencies, SONY was able to expand its business into new or related areas.

Meanwhile, thanks to a fertile imagination that brought many patent applications relating to user interfaces as well as articles in leading journals and presentations at international conferences, Jun Rekimoto started to become a poster boy of the age. Numerous other stars such as Toshiyuki Masui, François Pachet, Brian Clarkson, Ken Mogi, and Ivan Poupyrev also emerged to make their own active contributions, with numerous patents filed, papers published, and presentations given.

5.2 JIGSAW PUZZLES WITH PIECES MISSING AND ASSETS LEFT TO ROT

Even the best research is not guaranteed to generate saleable products. One reason for this is that, with the passage of time, it has become impossible to generate products based on only a single landmark new technology born from research results. In short, it is no longer possible to put together a final product without combining a diversity of technologies from different fields. The development of a high-tech TV, for example, requires not only a high-performance liquid crystal display panel but also technologies relating to backlighting, high-performance semiconductors, telecommunication and Internet capabilities.

Assuming the full range of technologies required for a desired product is available, a single research success can provide the final piece of a jigsaw puzzle. But in most cases, the opposite is true: The more cutting edge the research, the more pieces of the jigsaw seem to be missing. The time and work required mean that it is simply not cost effective for a single firm to plug all of these gaps themselves. To force such a project through might cause researchers to become dispirited at the sheer number of issues they need to address, then funding would eventually run out and the enterprise would be abandoned as a failure.

When dealing with leading-edge technologies, the approach of filling in all the gaps in the jigsaw puzzle within one's own company equates to the linear model described earlier. But in modern research and development, a single technology might have applications in a diversity of fields, and new products are born from combinations of technologies from entirely disparate disciplines. When seeking to produce next-generation products within the limited imaginative scope of a single company, the investment required to develop all of the necessary technologies is significant. And in addition to such costs, the narrow range of applications for the technologies developed means that potential returns are also likely to be greatly restricted.

What, then, is the correct approach? One should wait for almost everything to be ready before pushing ahead with product development. That is, one should wait until there are fewer empty spaces in the jigsaw puzzle. When seeking to commercialize a technology, one should intentionally plan for a time lag. From our own experience, we have found this lag to range from 5 to 10 years.

But that is not to say that fundamental research is unnecessary. Conducting fundamental research enables one to secure intellectual property rights and to maintain the relevant core competencies. And the leading position within the industry that results from such gains will in turn lead to profit. By giving up on fundamental research, one would forgo the opportunity to reap such rewards. It is a great shame that so many major firms have put an end to fundamental research and ceased investment in that area. Rather than abandoning fundamental research altogether, firms should conduct research according to an appropriate budget, while waiting for the appropriate peripheral technologies to be developed, either within their own organization or by other companies, before commencing attempts to create new products.

Fundamental research may take time, but it requires little in the way of funding. Technological development is more costly. Product development, meanwhile, demands investment on a massive scale—from the establishment of factories, to the procurement of parts and materials, to the establishment of supply and sales chains, and more besides. For manufacturers, fundamental research and technological development conducted on an appropriate scale will never incur costs that are potentially hazardous.

But, typically, 5 to 10 years down the line from a discovery, when the opportunity to commence product development is approaching, research staff will already have moved on to the next topic or even to the next after that. Most companies still adhere to this outdated model in which no time is set aside for the development of products from new discoveries. As a result, precious discoveries fall by the wayside, leading to frustration for firms as their rivals all enter into production of similar wares. Assets are left to rot. In order to prevent this, research management must determine the proper timing to proceed with product development.

6

From the Perspective of Technology Management

6.1 BUSINESS MANAGEMENT AND INNOVATION

In 1997, around 9 years after the establishment of Sony's Computer Science Lab (CSL), I left my post at Keio University and joined the SONY Corporation, later being appointed chief technology officer, in charge of technology management for SONY as a whole. Having for so long viewed research and development (R&D) from the perspective of a researcher, I now saw it from a business management perspective. For me, the opportunity to experience R&D from these two different standpoints was tremendous good fortune.

Here, let us look back on the relationship between business management and innovation. In *The Practice of Management*, Peter Drucker wrote: "Because the purpose of business is to create a customer, the business enterprise has two—and only two—basic functions: marketing and innovation. Marketing and innovation produce results...." There are various opinions as to the purpose of a business—from the creation of value, to the generation of profit. But according to Drucker, without customers, it is impossible to sell products. He asserts that this reality must be grasped as quickly as possible and an organic relationship established between the creation of customers through marketing and innovation (Figure 6.1).

"Marketing" is often interpreted as "market research," but Philip Kotler saw it instead as "market creation," writing: "The organization's marketing task is to determine the needs, wants, and interests of target markets and to achieve the desired results more effectively and efficiently than competitors, in a way that preserves or enhances the consumer's or society's well-being" (*Kotler on Marketing*).

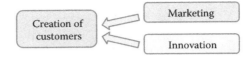

FIGURE 6.1
Business objectives. (Courtesy of Mario Tokoro/Sony Computer Science Laboratories, Inc., Tokyo, Japan.)

"Innovation," meanwhile, is often interpreted as "technological innovation." Joseph Schumpeter, however, described the term in conjunction with "entrepreneurship" thus: "(1) The production and carrying out of new products or new qualities of products, (2) the introduction of new production methods, (3) the creation of new forms of industrial organization (for instance, trustification) (4) the opening up of new markets, and (5) the opening up of new sources of supply" (*Economic Doctrine and Method: An Historical Sketch*). In other words, he views innovation as something that must create value not in the sense of short-term profits for individual companies but rather in terms of the potential to stimulate social and economic change. Examples of this include the steam engine, aircraft, electronics, computers, the Internet, and wireless communications technology such as mobile phones.

In each case, the time between the initial invention of these technologies and their eventual impact on society and the economy was anything between 30 and 100 years. The initial discovery is like completing the four sides of a jigsaw. As peripheral technologies are cultivated, subsequent pieces of the puzzle fall into place, and as these technologies are integrated, they take on the potential to benefit people's lives in a broad variety of ways. That is innovation in its truest form. And while it is not uncommon for the individual or company behind the initial invention to miss out on its benefits, there are also many cases in which firms have derived major success through expansion of their market facilitated by their early R&D efforts.

6.2 TECHNOLOGY: DEVELOPMENT PROCESS, TIME, AND COST

Here, I would like to take a broad look at the development process behind various technologies. The life cycle of a typical product runs from an "early days phase" that takes in fundamental research, through a "growth and development phase" during which technology is developed, and finally to the "maturity and decline phase." The product may be refined throughout the duration of its availability. Some technologies find themselves sidelined as alternatives emerge (Figure 6.2).

The fundamental research that takes place during the initial phase commonly relies on the creative capacity of individuals over a long period during which any kind of concerted planning is difficult. The research itself, however, is typically conducted by an individual or a small team and, as such, requires comparatively little funding: Computers, other experimental equipment, and the budget for a small number of overseas trips will generally be enough. And although exceptions come in the form of nuclear fusion, space exploration, and other so-called "big science," it is unusual for businesses to become engaged in such activities.

The growth and development phase is when products are produced based on newly discovered technologies. A closer examination, however, reveals that this phase can be divided into two distinct sub-phases. In the first sub-phase, the aim is to secure first-mover advantage by getting products onto the market as quickly as possible, typically in a period of around 3 years. As this period progresses, the emphasis shifts from simply creating a product to issues such as the costing of planning and production processes, or standardization with a view to product compatibility, and other issues similarly crucial to ensuring profitability over the longer term. It seems reasonable to anticipate that around 3 years might be

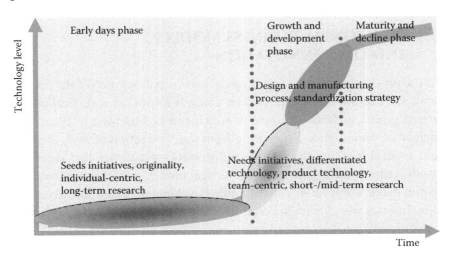

FIGURE 6.2
Research investment and management. (Courtesy of Mario Tokoro/Sony Computer Science Laboratories, Inc., Tokyo, Japan.)

spent on this second phase of development. Nonetheless, it is necessary to appreciate that commercialization strategies are likely to differ substantially between the two sub-phases of this growth and development phase.

During the growth and development phase and the phase of maturity and decline, significant costs are incurred in developing the technologies necessary to commercialize the product and on manufacturing. Such technological development is typically conducted within companies' own laboratories by a team 5 to 10 strong. However, processes such as the preparation of production facilities, the procurement of parts, the establishment of supply and sales chains, and sales planning mean that the actual commercialization of the product requires investment on a large scale. When dealing with mass production, this budget may amount to anything from 100 to 1,000 times as much as the initial fundamental research and 10 to several dozen times as much as technological development.

It is for this reason that even projects that received a green light up to and including the development of technology frequently find that, when it comes to full-scale product development, the answer is "no go." After all, major investment in an unripe market can pose a risk to the very survival of a business.

6.3 HORIZONTAL BUSINESS MODELS AND OPEN INNOVATION

At the production stage, a linear model in which every step of the process from start to finish is conducted in-house is known as a vertically integrated model, a classic example of which can be observed in the manufacture of automobiles. Everything from the production of steel, engines, and control systems through to semiconductor and product design is conducted in-house. Under such a vertical model, capital is concentrated in the major corporations that handle development and commercialization. In cases where firms successfully control the market prior to planned mass production, success can bring massive profits.

However, due to fierce competition in the sector, from 1990 onward it gradually became impossible for individual firms to invest in every aspect of R&D. Firms specialized in the production of semiconductors, computers, or electronic components, acquiring outstanding capability in each specific area. As a result, the vertically integrated model began to give way to a horizontally specialized one, with typical examples being mobile phone/smartphones and consumer electronics. The automobile industry now has a hybrid model as it is undergoing this shift.

Amid such change, conducting fundamental research in accordance with existing linear models grew more and more costly and less likely to bring profits. And almost exactly at the turn of the century, a new model of R&D emerged, the "open innovation" described by Henry Chesbrough (2003, *Open Innovation*). This new way broke free from the almost fundamentalist insistence among manufacturers on using technology developed in-house, a phenomenon sometimes referred to as "not invented here" or NIH. Under the open innovation approach, across the various phases—fundamental research, technology development, and product development—technologies from inside and outside the company are combined to accelerate product development (Figure 6.3).

Due to the unlimited scope for combining different technologies, it has become the norm for technology matching to take place online (*Wikinomics*, Don Tapscott, Anthony Williams, 2006). Open innovation is an approach to R&D for the Internet age.

Open innovation can be seen as a switch to a horizontal model of R&D. Conducting technology development on an increasingly specialized basis not only drives advances in such work, it also reduces the time and the budget required for product development. However, it is necessary to

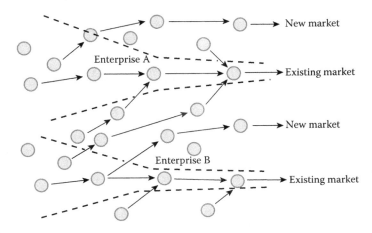

FIGURE 6.3
Open innovation. (Courtesy of Mario Tokoro/Sony Computer Science Laboratories, Inc., Tokyo, Japan.)

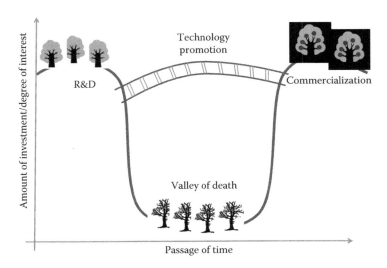

FIGURE 6.4
Valley of death (revised version). (Courtesy of Mario Tokoro/Sony Computer Science Laboratories, Inc., Tokyo, Japan.)

determine a level of compensation for the holders of various intellectual property rights that is in keeping with the product development process.

In addition, open innovation also makes it easier for enterprises to overcome the so-called "valley of death" (Figure 6.4). The depth of this valley depends on the number of individual technologies required for the development of a given product (in other words, the number of unfilled spaces in the jigsaw puzzle), while the width of the valley represents the time taken for the development of these technologies. Thanks to open innovation, the development of related technologies is conducted with a greater degree of transparency, boosting the probability that each will be utilized in an actual product and also reducing the time taken for this to occur.

6.4 MANAGEMENT OF OPEN INNOVATION

In this chapter, I have outlined the textbook definition of open innovation, along with the benefits of this approach. But when seen from the perspective of technology management in the actual running of a company, this approach is far from easy to put into practice. In order to make the most effective use of open innovation in product development, it is necessary to closely monitor trends in technology not only within one's own firm but also throughout the industry as a whole and also the world at large. Many jigsaw puzzles (representing potential future products) should be posited before the right timing is identified to procure appropriate technologies developed by other firms and to commence product development. Meanwhile, a decision can be made to sell to other companies the research conducted in one's own laboratory, in order to recoup the investment in technology development as quickly as possible so it can be reinvested in the development of next-generation technologies. For this, latent customers must be identified, while it is also possible to consider a strategy involving the implementation of marketing activities.

Whichever course of action is determined, it is crucial to emphasize the value of the research results that led to the development of the product in question both within one's own company and externally.

To execute all of these steps appropriately, it is first necessary to understand such aspects of the research or development at hand as its aims, nature, challenges faced, and so on, as well as to properly grasp the time and costs involved (Figure 6.5). Finally, it is necessary to maintain past

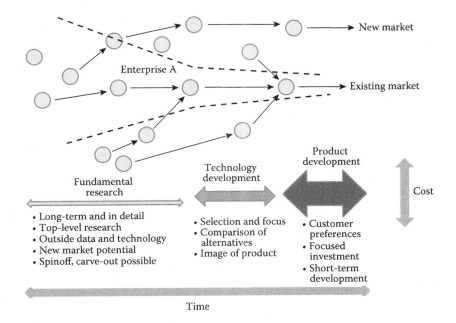

FIGURE 6.5
Management of open innovation. (Courtesy of Mario Tokoro/Sony Computer Science Laboratories, Inc., Tokyo, Japan.)

inventions, make improvements in line with the passage of time, and to constantly strive to bolster one's intellectual property rights.

From a technological perspective, it is not possible to traverse the valley of death simply through the investment of abundant money and personnel. Such instances require careful judgment of timing, making it vital to maintain a close watch on the status of the jigsaw while supporting further R&D. Above all, the marketing of the research that is at the heart of innovation is of paramount importance.

7

What TPO Represents

7.1 THE IMPLEMENTATION OF TPO

Backpedaling slightly, having waited for the correct timing, when it finally arrives and you assemble your product development team, what steps should be taken in order to ensure that your research results generate the anticipated value?

Back in 2004, there was an individual working concurrently at SONY and Sony's Computer Science Lab (CSL) who was uniquely gifted when it came to building interpersonal relationships; his name was Tsukasa Yoshimura. He came to see me after a business trip to Europe to tell me that commercialization of FourthVIEW was given a "no go."

"Well, we had yet to identify our market," I sensed. "And no matter how much we tried to push through suggestions for product development, it always seemed futile." At a time when Yoshimura was battling these frustrations, I went to him for advice: "The patents owned by Sony CSL are like buried treasure for SONY. I want to see them through to products for either SONY or other group companies. To achieve that I want to establish a Technology Promotion Office, which I hope can become the department in charge of marketing and sales of technology, and the production of products and services. Do you know of anyone with a passion for Sony CSL's technology, who could become a business producer, to do the rounds of SONY itself and other group companies, conduct research marketing and other marketing activities, and ultimately establish a product development team?"

"OK, let's try to find the right person," he replied. A few days later he returned with Tetsu Natsume in tow. I had encountered Natsume previously when he was working in Singapore as a specialist in charge of negotiations with government clients, and I remembered him especially because I enabled him to join Yoshimura's project. Speaking with him, I came to see how much the technologies of Sony CSL meant to him and his unrivaled passion for the development of products based on those technologies. I decided to invite Natsume to oversee our Technology Promotion Office (TPO).

But just as this momentous decision was reached, the then president of Sony CSL, Toshitada Doi, made it clear to me that although he would not oppose the move to establish TPO, he was not confident that it would produce the desired outcome. Furthermore, although there had been questions raised within the Research and Development (R&D) Management Section as to the need for TPO when we already had R&D Management, I myself was not aware of this discord. I earnestly pushed ahead with the establishment of TPO, and this was finally achieved in August 2004, with Natsume as manager and Yoshimura as an advisor, supported by two more members: Takahiro Sasaki and Yuji Ayatsuka.

With an intuitive grasp of my intentions for TPO, Natsume launched himself into cultivating customers, identifying the optimal timing to market the findings of our research, and grittily pushing ahead with the formation of teams for individual clients or integrated groups of clients. The successes this brought have already been described earlier in this book.

7.2 WHAT TPO REPRESENTS

Looking back, the establishment of TPO marked the change from a linear model in which a push was given to research successes to open innovation management that mediated between research and product development by linking the "push" from the laboratory with the "pull" of the product development division and managing the time lag accordingly. It is the promotion of technology that lies at the heart of innovation.

In the beginning, this marketing was primarily targeted toward SONY itself and other group companies, but nowadays, with major innovations in mind, external partners also feature in the equation. The establishment of TPO became the turning point in the shift from a linear model to an open innovation model of research management. Concrete examples of this were featured in Section I of this book.

As I said earlier, costs can generally be said to increase at an order of magnitude at each step from research to technology development to product development. My own experience of running Sony CSL bears that out. Viewed in reverse, all other things being equal, research must have a value 10 times greater than technology development and 100 times that of product development. The mission of Sony CSL is to pursue research firmly convinced of that value.

Furthermore, while it is the norm to lump research and development together as R&D, looking more closely at the characteristics of these two stages, research in its purest form is conducted without clear product development goals in mind, and the timescale before such research yields success is difficult to predict. In managing such research, it is therefore necessary to play the long game. In contrast, a clear idea of the eventual product that is integral to both technology and product development means that, at these stages, it is possible to identify a project's rivals, calling for great focus. I therefore believe that an approach under which these two stages are taken together as "technology and product development" offers the clearest sense of a project's timescale.

I imagine that most corporate laboratories are stuck in a loop of R&D. Sony CSL has been able to break free from this pattern and specialize in research, which can then be efficiently marketed thanks to the existence of TPO. Ultimately, this makes us an early adopter of open innovation management.

In Closing: Borderless Technology Promotion

To close, I'm here again (Tetsu Natsume) to talk about future developments in technology promotion.

Frankly, when I think about the future of technology promotion, I find it hard to imagine the direction it will take. I don't even know how research itself will change, or the world in general, so I feel there's nothing to do but keep on pressing ahead. It is also necessary for the approach to research marketing to keep on evolving, and I feel there will be a constant need for new methodologies to fit new research that emerges in the future.

Even so, there are several directions in which I would like to push the development of technology promotion in the long term. Among these, a key phrase is "act beyond borders." This represents the personal maxim of current Sony CSL President and Chief Executive Officer and Director of Research Hiroaki Kitano and is called for in every aspect of our research.

Research has long had an image of isolation from the wider world, with individual researchers cloistered in their ivory towers, unconcerned with anything beyond their own range of interest. In contrast, at Sony's Computer Science Lab (CSL), we encourage our researchers to deepen their research by personally going beyond borders and research disciplines to seek practical applications for their own findings.

For example, in the energy research described earlier, it was not simply a case of building a prototype energy server in the laboratory. The researchers themselves took the device to Africa and used it to power a public viewing of a FIFA World Cup match. The use of this technology in such a high-pressure atmosphere gave a real sense of what is really required in energy research.

Kitano himself has put these ideals of crossing borders into practice. From his original field of artificial intelligence and robotics, he moved on into biological sciences and, with this new perspective, established the new field of "systems biology."

Moving forward, I believe that this approach will be a crucial guiding principle for technology promotion.

TRANSCENDING NATIONAL BORDERS

For years, we have had a second laboratory in Paris and, at our Tokyo lab, there are also a great many researchers of different nationalities. And because English has long been the lingua franca of these laboratories, their activities are not limited to Japan. Our technology promotion activities also extend to Europe, the Americas, and Asia as part of a well-established global network.

Yet something that the energy project has made clear is the fact that technology in developing countries is heading in a different direction from that in industrialized nations. In what is known as "leap-frogging," emerging nations steal a march via a completely different developmental pathway from that of their industrialized counterparts, and we are seeing the spread of cutting-edge technology in developing nations.

Perhaps the best-known example of this is mobile phones. In industrialized nations, the advent of cell phones was after their wired predecessors became firmly established, and the two now share an infrastructure base. But the sudden spread of cell phone technology in developing nations has rendered the future development of a cable-based network highly unlikely.

In Africa, there is even a system for cash transfers to be made over the cellphone network using web money. In the absence of infrastructure such as automated cash machines, it seems that this is now the technology used by those working in the city who wish to send money back to their families in rural areas. With such systems yet to become widely available in industrialized nations, this early adoption of the latest technology in developing countries is one example of leap-frogging.

It seems likely that, in the absence of opposition from vested interests, such adoption of new technology instead of established infrastructures will see developing nations continue to move ahead of industrialized nations in certain technological fields. And information on such trends is also vital to our laboratories as a driver of fresh research.

TRANSCENDING INDUSTRIES

Although SONY is a global conglomerate with businesses in a diverse range of fields, including consumer electronics, computer gaming, finance, and entertainment, obviously there are still areas in which the company has no particular involvement. Therefore, in order to increase the likelihood that our research will find avenues for effective implementation, it is necessary to reach beyond the areas in which SONY is active and establish relationships in as wide a range of fields as possible. And one major benefit of the resulting, genre-spanning nature of our research is the way it helps to spawn new ideas.

For example, for several years, we have been engaged in an exciting collaboration with Daiwa House Industry, which has been probing potential crossovers between home-building and computer science. Daiwa House has know-how that we don't possess, such as how to open up new portals in walls for people to pass through. Our specialty, meanwhile, is the use of the latest software and compact electronic devices. It is exciting just to think what might be achieved by combining the know-how of both firms.

And by engaging in such collaborations not only with the housing industry but also in the fields of automobiles, food, toys, air travel, and more, there is limitless potential for new combinations.

TRANSCENDING GENERATIONS

For years, the technology promotion team has consisted of myself, Yoko Honjo, and Yoshiichi Tokuda, with Yoshimura in an advisory capacity. But, in 2015, with Sony CSL's research successes expanding, Honjo became CSL's general manager of general affairs and public relations. Tokuda meanwhile, parallel to his role with the Technology Promotion Office or TPO, became the manager of the Open Energy Project. But to make sure the team's activities are passed on to the next generation, 26-year-old Kojiro Kashiwa was brought onboard. Born in 1988, the very year CSL was founded, he represents the next generation, and there are great hopes for the future developments of this new, epoch-spanning stage of technology promotion.

TRANSCENDING SONY CSL

Up to now, our research marketing has been focused on the research of Sony CSL. But in the belief that there are countless researchers out there in need of the sort of service technology promotion is able to provide, we are hoping to take our activities beyond the confines of this company.

Afterword

For me, as one of the cowriters of this book, the idea for the volume came from a conversation, at a reception of the lab's open house 5 years ago, with Shinko Yuri of Sci-Tech Communications. Yuri had cowritten a book with Mario and had interviewed me about the Technology Promotion Office (TPO). My account of TPO's work had stuck with her, and she told me how fascinating it was to encounter something so unusual, even compared to other research institutions.

Thinking she was onto something, I decided to compile details of our TPO work but, with no previous experience as a writer, I was unsure how to go about it. In the end, though, thanks to the guidance of others, things began to take shape.

While tackling this book and digging back through our work to date, I have been able to get a clearer overview of what we have been doing. I believe that we were able to clearly present the methodology behind research management, and the origins of this enterprise, along with its significance. And interviewing our various researchers also bore valuable fruit as I gained insights into the importance of research management from their perspective.

This time, I was chosen to present the story of TPO, but the real bedrock of the Research Management Team's activities over the last decade has been the first-generation members Sasaki and Ayatsuka, second generation Honjo and Tokuda, and of course Yoshimura, who not only provided the initial impetus for TPO, but who has since remained in an advisory capacity behind the scenes. I owe them all my heartfelt gratitude.

I must also express my thanks to company CEO Hiroaki Kitano, both for giving me this opportunity and for spurring me on once the decision to produce this book had been taken.

And of course, this book would never have been possible without our founder, Mario Tokoro, who read my drafts, discussed the finer points with me, and provided all sorts of invaluable guidance. He too is owed my heartfelt gratitude.

Finally, I would like to thank Yuri—for providing the initial inspiration for this book as well as providing guidance from the very first steps in the process—along with Katayose and Fukuda for patiently editing and correcting my drafts. I would also like to express my appreciation both to Adam Fulford, for overseeing the translation of the book into English, and to Francois Pachet, who did so much to contribute to the publication of the English version.

Tetsu Natsume

Back when I first asked Tetsu Natsume to develop and implement TPO, I cannot say I had a clear idea of how the enterprise would be operated. From that beginning, I watched him gradually move from feeling his way along to dynamically promoting our research, striving to transform intellectual property into products and, with more than a few failures along the way, developing a clear model for technology promotion, the results of which are presented in this book. I am sure that without him it would have been impossible for technology promotion to come this far. And by observing his activities, I too have gained a deeper understanding of the laboratory environment and have learned a great deal. I am extremely grateful.

In Section III, I re-presented many of the examples given in Section I from a technology-management perspective. That's how I feel I was able to clearly explain such concepts as getting past the valley of death and open innovation. As well as helping to understand the fundamentals of technology management and research marketing, I will be glad if this text can be of some practical use.

My thanks go once more to Tetsu Natsume as well as my heartfelt gratitude also to Yuri, Katayose, and Fukuda for working on tight deadlines to produce a very readable text. And last but not least, I would like to thank Adam Fulford for overseeing the translation of the book into English.

Mario Tokoro

Index

A

Academic partners, 64–65
Academic research, 103–106
Advertising, 11
Africa, 128
Aha! Experience Project, 53–58, 66–71
AIBO, 15, 24, 25
AMInet Project, 112
Andre, Alexis, 17, 101–103
Apertos project, 112
Appliance anthromorphization, 40–43
AR. *See* Augmented reality
Asperios, 24
Augmented reality (AR), 15
 invention of, 7–9
 progress of, 12
Auto-completion, 18
Ayatsuka, Yuji, 9, 25

B

Barcode technology, 9–10
Battery technology, 65–69
Beta testing, 45–47
Big science, 117
Blue-sky research, 7–8
Bluetooth, 18, 20, 37
Borderless technology promotion, 127–129
Business management, 115–116
Business models, horizontal, 118–120

C

Case studies
 12Pixels, 43–47
 Aha! Experience Project, 53–58
 CSL Paris, 38–40
 econophysics, 49–53
 EDS, 38–40
 interindustry collaborations, 40–43
 location-sensing service, 58–63

Moe-Kaden, 40–43
Open Energy Systems Project, 63–74
VAIO Pocket, 29–31
Chesbrough, Henry, 119
Clarkson, Brian, 20
Collaborative research, 92, 105
Commercial applications, 13, 22–23, 113–114
Corporate partners, 64–65, 105
Corporate research, 103–106
CROSS YOU, 35–36
CSL Paris, 17, 38–40
CyberCode, 9–10, 11

D

Daiwa House, 42–43, 105, 129
Demo road shows, 83–85, 104
Developing nations, 128
Development process, 116–118
Digital appliances, 40–42
Digital cameras, 8
Distributed object-oriented OS, 112
Doi, Toshitada, 111, 124
Drucker, Peter, 115

E

Econophysics, 16, 49–53
EDS. *See* Extractor Discovery System
Electric power industry, 63–74
E-mail newsletter, 85–88
Entertainment media, 53–58
ERATO. *See* Exploratory Research for Advanced Technology (ERATO) program
Exploratory Research for Advanced Technology (ERATO) program, 15
Extractor Discovery System (EDS), 38–40
EyeToy, 9

133

F

Feature phones, 43–47
FEEL, 19–20, 31–38
    history of, 38
    implementaion on mobile phones, 35–36
    as landmark idea, 31–33
    NFC standard adoption, 34–35
    smartphones and, 36–37
    videoconferencing system, 33–34
    workings of, 32–33
FeLiCa, 20, 32, 34, 35–36
Flower Robotics, 15
FourthVIEW, 14
Fundamental research, 114, 117, 119

G

Galapagos, 44–45
Geometric sciences of information, 16–17
Gesture recognition technology, 20–21, 40
Ghana, 66–68, 69–71
G-sense, 30

H

Head-mounted display (HMD), 79
Honjo, Yoko, 12, 61–63, 129
Horizontal business models, 118–120
Hoshino, Masaaki, 39
Human-computer interaction, 15

I

Information-gathering channels, 78–79
Innovation
    business management and, 115–116
    open, 118–122
Interindustry collaborations, 40–43
Internet, 8
Internet of electricity, 64, 69, 71–74
Intuitive devices, 20–21
Isozaki, Takashi, 17, 106–108

J

JackIn, 79
Jello printer, 41

K

Kasahara, Shunichi, 79
Kitano, Hiroaki, 15, 51, 127
Kitano Symbiotic System Project, 15
Kobayashi, Yoshiyuki, 39–40
Kojima, Tamaki, 39
Koozyt, Inc., 60–63
Kotler, Philip, 115

L

Lab-driven product development, 111–113
Leap-frogging, 128
License fees, 91–92
Lifelog, 20, 21
Lithium-ion (Li-ion) batteries, 65–66
Location-sensing service, 58–63

M

Market
    bringing research to, 11–12
    difficulty of bringing technical breakthroughs to, 31–38
    readiness of, 11–12, 97
Marketing, 115–116
Masui, Toshiyuki, 18
Memorandum on Technology Release (MoTR), 35, 36, 88–91
Mental fortitude, 95
Miyaki, Kazuhito, 9
Miyaki, Satoru, 9
Mobile phones, 128
    charging service, for unelectrified areas, 69–71
    FEEL-enabled, 35–36
Moe-Kaden, 40–43
Mogi, Ken'ichiro, 16, 53–58
Music categorization technology, 38–40

N

Natsume, Tetsu, 123–124
Near-field communication technology, 20, 32–35; *see also* FEEL
NeWS Workstation, 111
Next-generation electrical infrastructure, 63–74

Nielsen, Frank, 16–17
Not invented here (NIH), 119

O

Okinawa Institute of Science and
 Technology Graduate University
 (OIST), 64–65, 71–74
One-Touch, 36–37
One-touch operation technology, 18–20
Open Energy Systems (OES) Project,
 63–74, 129
Open innovation, 118–122
Open-source software, 112
Owada, Shigeru, 17, 40–42

P

Pachet, François, 17, 21, 38
Packetized electric power, 65
Personal computers (PCs), 8
Personal connections, 53
Pixel art, 43–47
PlaceEngine, 58–60, 62
Place project, 58–63
PlayStation (PS), 8
PlayStation Portable (PSP), 54
POBox, 18–20, 24–25, 27
Poupyrev, Ivan, 44
Predictive text input, 18, 24
Presense technology, 30–31
Product development
 lab-driven, 111–113
 life cycle, 116–117
 process, 116–118
Product development cycle, 36–37
PS2, 9
Public betas, 45–47

Q

Qualia, 16

R

Real-time protocol, 112
Rekimoto, Jun, 8, 9, 11, 15–16, 19–20,
 30–31, 37, 58, 61, 103–106
Renewable energy, 71–72

Research
 academic vs. corporate, 103–106
 blue-sky, 7–8
 bringing to market, 11–12
 collaborative, 92, 105
 commercial applications of, 13, 22–23,
  113–114
 fundamental, 114, 117, 119
 impact of selling on, 99–100
 at Sony CSL, 12–13, 21–22, 24–25
 systems, 49
 theoretical, 49–51
Research and development (R&D), 14
 gap between commercialization
  and, 22–23
 SONY corporate, 39–40
Research discoveries, cataloging
 of, 75–79
Researchers
 communication with, 79
 role division between technology
  promoters and, 98
 on technology promotion, 101–108
 TPO interview of, 76–78
Research planning, 99
Review talk, 75–76, 77

S

Sakurada, Kazuhiro, 16
Sales collateral, 80–82
Sales sheet, 80–81
Sasaki, Takahiro, 25–26
SBG. *See* Semiconductor Business Group
Schematized FAQs, 81–82
Schumpeter, Joseph, 116
Sega Corporation, 54–58
Semiconductor Business Group (SBG),
 51–52
SEND. *See* Sony Energy Device
Setting divergences, 97
Shionozaki, Atsushi, 58, 61, 112
SmartBand, 20
Smart Operation, 20–21, 40
Smartphones, 11, 15–16, 18–20, 36–37,
 40, 44
So-net, 60
Sony business units, 22–23, 27–28, 36
Sony Computer Entertainment (SCE), 9, 57

Sony Computer Science Labs (Sony CSL)
   blue-sky research by, 7–8
   changing research style of, 24–25
   crediting of, 27–28, 30–31, 34–37, 88, 90–91
   founding of, 111–113
   interindustry collaborations, 40–43
   license fees and, 91–92
   obscurity of, 23–25
   research at, 12–13, 21–22, 49
   researchers at, 15–18, 22, 23
   review talk, 75–76, 77
   spin-off, 58–63
   technology derived from, 18–23
   Technology Promotion Office, 3–28
   transcending, 129
   vision of, 14–15
Sony Energy Device (SEND), 65–66, 67, 69
Sony Ericsson, 46–47
Sony Music Artists (SMA), 54–55, 57
Spin-off, 58–63
Subtropical and Island Energy Infrastructure Technology Research Subsidy Program, 72
Sueyoshi, Taka, 58, 60, 61
Systems research, 49

T

Taiken Kukan, 45
Tajima, Shigeru, 65
Takayasu, Hideki, 16, 50, 51–52
Technical breakthroughs, difficulty of bringing to market, 31–38
Technology features, 80
Technology management, 115–122
Technology on Promotion (TOP) list, 77–78
Technology promotion
   borderless, 127–129
   cataloging of research discoveries, 75–79
   core principles of, 92–96
   future of, 127
   MoTR and, 88–91
   multiple information-gathering channels for, 78–79
   need for, 96–100
   as part of lab, 98
   researchers on, 101–108
   role of, 7–12, 13
   sales collateral development, 80–82
   selling, 82–88
   status of, 98–99
   techniques for, 75–100
   TPO interview, 76–78
   workflow, 76
Technology Promotion Office (TPO)
   about, 12–14
   action plan of, 26–28
   creation of, 25
   functions of, 7–12, 13
   history of, 111–125
   implementation of, 123–124
   interview, 76–78
   mission of, 25–26
   motto of, 93
   next-level challenges for, 49–74
   precursors to, 111–114
   representativeness of, 124–125
   sales tools, 82–88
   staff of, 12
   typical workday in, 3–7
Technology summary, 80
Technology tags, 81–82
Technology transfer, 24–25
   case studies in, 29–47
   credit and, 88, 90–91
   failures, 29–30
   follow-up after, 96
THE EYE OF JUDGMENT™, 10–11
Theoretical research, 49–51
Timing gaps, 11–12, 97
Tokoro, Mario, 13–14, 15, 33, 51, 60, 63, 73–74
Tokuda, Yoshiichi, 12, 69–71, 74, 129
Tokyo National Museum, 61–63
TOP. *See* Technology on Promotion (TOP) list
Touch screens, 45
T-pop News, 85–88, 104
Tsunakawa, Toyohiro, 51
12Pixels, 43–47

V

VAIO-C1, 9
VAIO Pocket, 29–31

Valley of death, between research and commercialization, 22–23, 120
Vaporware, 29
Videoconferencing system, 33–34
Video games, 9–10

W

Watanabe, Yusuke, 9
Wi-Fi, 20, 32, 37, 58–59
Willis, Karl, 44
Wireless devices, 31–32, 37
World Cup, 66–68, 69

X

Xperia smartphones, 18–20, 40

Y

Yasuda, Masashi, 66
Yokote, Yasuhiko, 112
Yoshimura, Tsukasa, 9, 12–14, 24–25, 51, 66–68, 123–124

Z

ZMP, 15